AF849478

Governance and Planning of Mega-City Regions

Neoliberalism's market revolution has had a tremendous effect on contemporary mega-city regions. The negative consequences of market-oriented politics for territorial growth have been recognized. While a lot of attention has been given to how planners and policymakers are fighting back political fragmentation through innovative governance and planning, little has been done to reveal such practices through an international comparative perspective.

Governance and Planning of Mega-City Regions provides a comparative treatment and examination of how new approaches in governance and planning are reshaping mega-city regions around the world. The contributors highlight how European mega-city regions are evolving and how strategic intervention is being redefined to enable the integration of urban qualities in a multi-level governance environment; how traditional federal countries in North America and Australia see the promise of major policies and development initiatives finally moving ahead to herald a more strategic intervention at national and regional scales; and how transitional economies in China witness the rise of state strategies to control the articulation of scales and to reassert the functional importance of state in a growing diffused power context.

This book offers case studies written from a variety of theoretical and political perspectives by world-leading scholars. It will appeal to upper level undergraduates, postgraduates, researchers, and policymakers interested in urban and regional planning, geography, sociology, public administrations, and development studies.

Jiang Xu is Assistant Professor in the Department of Geography and Resource Management, the Chinese University of Hong Kong. She is an urban and regional specialist and co-author of *Urban Development in Post-Reform China: State, market and space* (Routledge 2007).

Anthony G.O. Yeh, Academician of the Chinese Academy of Sciences, is Chair Professor and Head of the Department of Urban Planning and Design and Director of the Centre of Urban Studies and Urban Planning of the University of Hong Kong. His main areas of specialization are in urban development and planning in Hong Kong and China, and the applications of GIS in urban and regional planning.

Routledge Studies in Human Geography

This series provides a forum for innovative, vibrant, and critical debate within Human Geography. Titles will reflect the wealth of research that is taking place in this diverse and ever-expanding field. Contributions will be drawn from the main sub-disciplines and from innovative areas of work that have no particular sub-disciplinary allegiances.

Published:

1 **A Geography of Islands**
Small island insularity
Stephen A. Royle

2 **Citizenships, Contingency and the Countryside**
Rights, culture, land and the environment
Gavin Parker

3 **The Differentiated Countryside**
Jonathan Murdoch, Philip Lowe, Neil Ward and Terry Marsden

4 **The Human Geography of East Central Europe**
David Turnock

5 **Imagined Regional Communities**
Integration and sovereignty in the global south
James D. Sidaway

6 **Mapping Modernities**
Geographies of Central and Eastern Europe 1920–2000
Alan Dingsdale

7 **Rural Poverty**
Marginalisation and exclusion in Britain and the United States
Paul Milbourne

8 **Poverty and the Third Way**
Colin C. Williams and Jan Windebank

9 **Ageing and Place**
Edited by Gavin J. Andrews and David R. Phillips

10 **Geographies of Commodity Chains**
Edited by Alex Hughes and Suzanne Reimer

11 **Queering Tourism**
Paradoxical performances at Gay Pride parades
Lynda T. Johnston

12 **Cross-Continental Food Chains**
Edited by Niels Fold and Bill Pritchard

13 **Private Cities**
Edited by Georg Glasze, Chris Webster and Klaus Frantz

14 **Global Geographies of Post Socialist Transition**
Tassilo Herrschel

15 **Urban Development in Post-Reform China**
State, market and space
Fulong Wu, Jiang Xu and Anthony Gar-On Yeh

16 **Rural Governance**
International perspectives
Edited by Lynda Cheshire, Vaughan Higgins and Geoffrey Lawrence

17 **Global Perspectives on Rural Childhood and Youth**
Young rural lives
Edited by Ruth Panelli, Samantha Punch, and Elsbeth Robson

18 World City Syndrome
Neoliberalism and inequality in Cape Town
David A. McDonald

19 Exploring Post Development
Aram Ziai

20 Family Farms
Harold Brookfield and Helen Parsons

21 China on the Move
Migration, the state, and the household
C. Cindy Fan

22 Participatory Action Research Approaches and Methods
Connecting people, participation and place
Sara Kindon, Rachel Pain and Mike Kesby

23 Time-Space Compression
Historical geographies
Barney Warf

24 Sensing Cities
Monica Degen

25 International Migration and Knowledge
Allan Williams and Vladimir Baláž

26 The Spatial Turn
Interdisciplinary perspectives
Barney Warf and Santa Arias

27 Whose Urban Renaissance?
An international comparison of urban regeneration policies
Libby Porter and Katie Shaw

28 Rethinking Maps
Martin Dodge, Rob Kitchin and Chris Perkins

29 Rural–Urban Dynamics
Livelihoods, mobility and markets in African and Asian Frontiers
Jytte Agergaard, Niels Fold and Katherine V. Gough

30 Spaces of Vernacular Creativity
Rethinking the cultural economy
Tim Edensor, Deborah Leslie, Steve Millington and Norma Rantisi

31 Critical Reflections on Regional Competitiveness
Gillian Bristow

32 Governance and Planning of Mega-City Regions
An international comparative perspective
Jiang Xu and Anthony G.O. Yeh

Not yet published:

33 Design Economies and the Changing World Economy
Innovation, production and competitiveness
John Bryson and Grete Rustin

34 Globalization of Advertising
Agencies, cities and spaces of creativity
James Faulconbridge, Peter J. Taylor, J.V. Beaverstock and C. Nativel

35 Cities and Low Carbon Transitions
Harriet Bulkeley, Vanesa Castán Broto, Mike Hodson and Simon Marvin

36 New Economic Spaces in Asian Cities
From industrial restructuring to the cultural turn
Peter W. Daniels, Kong Chong Ho and Thomas A. Hutton

First published 2011
by Routledge
2 Park Square, Milton Park, Abingdon, Oxon OX14 4RN

Simultaneously published in the USA and Canada
by Routledge
270 Madison Avenue, New York, NY 10016

Routledge is an imprint of the Taylor & Francis Group, an informa business

© 2011 Selection and editorial matter, Jiang Xu and Anthony G.O. Yeh;
individual chapters, the contributors

Typeset in Times New Roman by
Florence Production Ltd, Stoodleigh, Devon, UK
Printed and bound in Great Britain by
CPI Antony Rowe, Chippenham, Wiltshire

All rights reserved. No part of this book may be reprinted
or reproduced or utilized in any form or by any electronic,
mechanical, or other means, now known or hereafter invented,
including photocopying and recording, or in any information
storage or retrieval system, without permission in writing
from the publishers.

British Library Cataloguing in Publication Data
A catalogue record for this book is available from the British Library

Library of Congress Cataloguing in Publication Data
Governance and planning of mega-city regions: an international
comparative perspective/edited by Jiang Xu and Anthony G.O. Yeh.
p.cm.
Includes bibliographical references and index.
1. City planning—Cross-cultural studies. 2. Regional planning
—Cross-cultural studies. I. Xu, Jiang, 1968–
II. Yeh, Anthony G. O., 1952–
HT166.G663 2010
307.1'2164–dc22 2010005147

ISBN: 978–0–415–56089–4 (hbk)
ISBN: 978–0–203–86551–4 (ebk)

Governance and Planning of Mega-City Regions

An international comparative perspective

Edited by
Jiang Xu and
Anthony G.O. Yeh

Routledge
Taylor & Francis Group
NEW YORK AND LONDON

Contents

Illustrations

Figures

Tables

Contributors

John Abbott is a practicing Metropolitan Planner in South East Queensland, Australia. He was previously the Project Coordinator of the SEQ 2001 and SEQ 2021 regional planning projects. He teaches planning theory and metropolitan planning at the University of Queensland. He has analyzed metropolitan planning processes in South East Queensland, Greater Vancouver, and New York using concepts of planning as managing uncertainty.

Louis Albrechts is Professor of the Department of Architecture, Urbanism and Planning at the Catholic University of Leuven, Belgium. His research interests include strategic spatial planning, sustainable development, and regional design, and he has published widely on these issues. He is the founder and co-editor of *European Planning Studies*, a corresponding member of the German Academy for Research and Planning, and a member of the Advisory Board of the global Research Network on Human Settlements.

Brian Cheers is Research Professor Emeritus of Community Development and former Director of the Center for Rural and Regional Development at the Whyalla Campus of the University of South Australia, Australia. He is also Founding Director of the Northern Australia Research Institute and the Center for Social and Welfare Research at James Cook University. He has published four books and many monographs and papers on rural and regional issues.

Lu Chen was a student of the Department of Urban and Regional Planning at Nanjing University, China.

Chaolin Gu is Professor of the School of Architecture, Tsinghua University, Beijing, China. He has published sixteen books and over 260 articles on urban and regional planning, regional economics, and urban geography in China. He is the principal investigator of a number of projects on China's urban and regional development and planning. He is Vice President of the Chinese Geographical Association, and serves on the editorial boards of many journals and academic councils.

Peter Hall is Bartlett Professor of Planning and Regeneration at the Bartlett School of Architecture and Planning, University College London, UK. He has received the Founder's Medal of the Royal Geographical Society for distinction in research, and is an honorary member of the Royal Town Planning Institute, which awarded him its Gold Medal in 2003. He holds fourteen honorary doctorates from universities in the UK, Sweden, and Canada. He received the 2005 Balzan Prize for work on the Social and Cultural History of Cities since the Beginning of the 16th Century. He is a Fellow of the British Academy and the European Academy and President of the Town and Country Planning Association. He was knighted in 1998 and in 2003 was named by Her Majesty Queen Elizabeth II as a "Pioneer in the Life of the Nation" at a reception in Buckingham Palace.

Jim Harvey is Adjunct Professor of the Center for Rural Health and Community Development at the University of South Australia, Australia. His most recent publications have been on intra-regional cooperation in urban and regional development. He is currently the Australian Manager of an Australian Aid (AusAid) community development project in the Eastern Highlands Province of Papua and New Guinea.

Linda McCarthy is Associate Professor of Geography and Urban Studies at the University of Wisconsin-Milwaukee, USA, and is also a certified planner. Her research focuses on urban and regional economic development and planning in the United States, Western Europe, and China. Her recent publications have been on regional cooperation instead of wasteful competition for corporate investment; government subsidies for automobile plants; environmental justice and brownfield redevelopment; and the globalization of the economy.

Willem Salet is Professor of Urban and Regional Planning in the Faculty of Social and Behavioral Sciences at the University of Amsterdam, The Netherlands. He is also the President of the Association of European Schools of Planning. His research specializes in spatial planning and metropolitan governance, urban networks, and decision making in strategic urban projects. He coordinated various research projects on behalf of the European Union, national ministries, the National Scientific Council for Government Policy, and other stakeholders in the field of urban studies, and has published widely on regional planning and governance.

Saskia Sassen is Robert S. Lynd Professor of Sociology in the Department of Sociology and Member of the Committee on Global Thought at Columbia University, USA (www.saskiasassen.com). Her most recent books are *Territory, Authority, Rights: From Medieval to Global Assemblages* (2008) and *A Sociology of Globalization* (2007). Her books have been translated into twenty-one languages. Her comments have appeared in numerous newspapers, and she regularly writes for www.HuffingtonPost.com and www.OpenDemocracy.net. She serves on several editorial boards and is an advisor to several international bodies. She is a member of the Council on Foreign Relations, a member of the National Academy of Sciences Panel on Cities, and chaired the Information Technology and International Cooperation Committee of the Social Science Research Council (USA).

Chun Wang is an urban planner in the Master Planning Department of Tsinghua Urban Planning and Design Institute, Beijing, China.

Jiang Xu is Assistant Professor in the Department of Geography and Resource Management, the Chinese University of Hong Kong, Hong Kong SAR. She is a specialist in urban and regional issues, and is currently leading research projects in intercity competition and cooperation, as well as urban and regional governance in China. She has published widely on urban and regional development in leading international journals and is co-author with F. Wu and Anthony G.O. Yeh of *Urban Development in Post-Reform China: State, market and space* (Routledge 2007). She was the recipient of the 2008 Research Output Prize of the University of Hong Kong.

Robert Yaro is President of Regional Plan Association, America's oldest independent metropolitan policy, research, and advocacy group. He is also Professor of Practice in City and Regional Planning at the University of Pennsylvania, USA. He has taught at Harvard University and the University of Massachusetts. He co-chairs the Empire State Transportation Alliance and the Friends of Moynihan Station, and is Vice President of the Forum for Urban Design. He serves on Mayor Bloomberg's Sustainability Advisory Board, which helped prepare PlaNYC 2030, New York City's new long-range sustainability plan.

Anthony G.O. Yeh is Academician of the Chinese Academy of Sciences. He is also Chair Professor and Head of the Department of Urban Planning and Design and Director of the Centre of Urban Studies and Urban Planning at the University of Hong Kong, Hong Kong SAR. His main areas of specialization are in urban development and planning in Hong Kong and China, and the application of GIS in urban and regional planning. At present, he is Secretary-General of the Asian Planning Schools Association and Asia GIS Association. He is on the editorial boards of key international and Chinese journals, and has published over thirty books and monographs, and over 180 academic journal papers and book chapters. He received the 2008 UN-HABITAT Lecture Award for his outstanding and sustained contribution to research, thinking, and practice in the human settlements field.

Taofang Yu is Lecturer of the School of Architecture at Tsinghua University, Beijing, China. He has published four books and about fifty articles on urban competition and the mega-city region.

Cheng Zhang is a certified Urban Planner at the Nanjing Urban Planning Bureau, China. He has published five articles on the mega-city region and the mega-project.

Min Zhang is Associate Professor of the Department of Urban and Regional Planning at Nanjing University, China. She has published about thirty articles on urbanization, the megalopolis, and the global city region.

Xiaoming Zhang is a PhD candidate in the School of Architecture at Tsinghua University, Beijing, China. He has published over six articles on the mega-city region and spatial regionalization.

Acknowledgments

This book originated from the International Symposium on Mega-City Regions: Innovations in Governance and Planning held at the Centre of Urban Studies and Urban Planning of the University of Hong Kong in August 2008. The symposium was sponsored by the Zhong Hua Real Estate and Construction Research and Development Foundation and Professor Francis S.K. Lau.

Jiang Xu is most grateful to the General Research Fund of the Research Grants Council of Hong Kong (Project code: CUHK453609) for supporting the research.

Anthony Yeh would like to thank Mrs. Li Ka Shing Fund and Strategic Research Theme in Contemporary China Studies of the University of Hong Kong for supporting the research.

1 Governance and planning of mega-city regions

Diverse processes and reconstituted state spaces

Jiang Xu and Anthony G.O. Yeh

Introduction

Cities are increasingly at the center of global production and consumption as well as social and political transformation. Over the past several decades, their role as important nodes of global networks of commercial, social, and cultural transactions has expanded, creating new types of sprawling, often multi-centered, urban agglomeration. Various labels have been employed to describe this phenomenon of large-scale urbanization, such as the metropolis, the conurbation, the megalopolis, and the global city region (Sassen 2001; Scott 2001). The idea of the "mega-city region" represents a new phase of development of this theoretical lineage to designate a novel spatial format. As defined by Hall and Pain (2006), a mega-city region is a cluster of contiguous cities or metropolitan areas that are administratively separate but intensively networked, and clustered around one or more larger central cities. These places exist both as separate jurisdictional entities, in which most residents work locally, and as parts of a wider functional urban region connected by dense flows of people and information (Hall and Pain 2006). In this sense, the mega-city region represents a "new scale of urban agglomeration" because of its highly polycentric nature, its wide geographical coverage, and its privileged position in the global economy.

Mega-city regions constitute distinctive regional spatial formations that are undergoing major transformation due to the impact of globalization. They are at the forefront of this transition, and many have become national engines of economic growth. Globally, there are over 450 (mega) city regions with more than 1 million residents; at least twenty of them have populations of over 10 million (Scott 2001; United Nations 2004). These regions house growing populations in relatively small land areas. Some expand even faster into urbanized locations of over 50 million inhabitants, as evidenced in developing countries such as China. Their footprint will have a direct impact on environmental change, land use pattern, and spatial transformation, as well as on the lives of existing and new city dwellers alike.

Regions, no matter what they are called, are becoming highly complex and functional entities. Allen Scott (2001) contemplates that city regions (e.g. metropolitan agglomerations and polycentric urban networks), not central cities or nation-states, seem to be the imposing scales for competition in a world economy under global capitalism. More than ever, national and local economic policies have to be linked more directly to diverse forms of spatial planning, infrastructure investment, and socio-economic governance on a regional scale. The intensification of this "city regionalism" view (Jonas and Ward 2007) in recent years signifies a critical alteration of the state capitalism in the Fordist–Keynesian accumulation regime and the neoliberal urbanism in the post-Fordist era. Such a shift carries a strong implication that the region has increasingly become an important spatial scale for state regulation, capital accumulation, and political compromise in the global north.

Relative to the regions of the advanced capitalist states, we know much less about how regions have evolved in developing countries, albeit the largest and fastest-growing urbanized locations are situated in this part of the world. However, a cursory glance at what is happening in these countries would seem to indicate an intensifying interest in canonizing the regional scale as a functional space for economic planning and political governance. The super-agglomeration, or city regions, in the Third World countries have attracted substantial attention. A growing number of scholars have sought to map the emergence of this new scale of political economic activities and to demonstrate how such profound reshufflings represent key moments in an intense struggle to compete in the competitive global capitalism (e.g. Scott 2001; Stren 2001; Douglass 2001, 2002; Laquian 2005; Wu and Zhang 2007; Xu 2008). Primarily referring to China's mega-city regions, Xu (2008) demonstrates how these regions are seen as important structural and strategic expressions in the post-reform transition of repositioning places in the national and global economic landscapes.

The notion of regional resurgence as political and economic organization requires a careful reading. Many theoretical interpretations in both the developed and the developing world have been built upon a rather narrowly set perception to imply that the emergent superimposition and increasing socio-economic interdependencies at the regional scale are mainly a major response to globalization and interspatial competition. Viewed in this way, regions are often examined in terms of their functionality for regulation, development, and competition within a globalizing economy. They are theorized as the architectural, social, cultural, and spatial building blocks of the global economy (Jonas and Ward 2007). Such an interpretation is provoking much admirable work to proclaim the rise of regions as a scale of wealth creation and socio-political formation (e.g. Ohmae 2001; Porter 2001; Scott et al. 2001; Scott and Storper 2003; Thierstein and Förster 2008). However, the interpretation may show negligence. It lacks a serious treatment of city regionalism as a differentiated socio-political process constructed around various places, spaces, and scales in a context-sensible way. This amounts to a risk of applying a hegemonic discourse to understand globalization and city-regionalism at the expense of knowledge about place-specific practices. Indeed, a careful reading of these practices could prove most valuable in establishing a meaningful and insightful sense of the contextually charged relations behind the reconstituted state and the rise of new state spaces (Swyngedouw 1997; MacLeod 1999, 2001; Brenner 2002). Our overall purpose, then, is to rediscover different mega-city regions in different circumstances by treating them not only as functional and competitive nodes of global capitalism but also as products of diverse processes and contextually reconstituted state spaces.

In this sense, our perspective based on case studies and comparative interpretations is particularly insightful. Brenner et al. (2003) deem the growing importance of regional space to be punctuated by several inadequacies and missing links in academic inquiry and political discourse. Among these are: (1) a lack of case studies to differentiate interpretations of historical and contemporary state spaces in diverse contexts, territories, and scales; and (2) a reluctance to conduct comparative studies that shed light on the divergent pathways in and through which state space is reorganized, and on their often markedly different outcomes in social, political, institutional, and geographical terms. It is this evocative commentary that prompts us to deploy the perspective of this volume.

Our general observation is that while mega-city regions in different parts of the world share a commonality—rapid transformation in an era of globalization—they also have different rationales, development patterns, fiscal capacities, managerial abilities, and experiences in regional governance and planning (Vogel 2010). In addition, they are evolving in diverse political contexts and economic landscapes. The roles of their public and private sectors in regional formation vary in form and sophistication (Vogel 2010). Although much work on (mega-)city regions now exist (e.g. Simmonds and Hack 2000; Hall 2001; Scott 2001; Herrschel and Newman 2002; Salet et al. 2003; Laquian 2005; Hall and Pain 2006; Kidokoro et al. 2008), none is devoted to explore experiences and broad questions related to the governance and planning in mega-city regions from an international comparative perspective. With case studies written from a variety of theoretical and political perspectives, this volume brings together a group of world-leading scholars to draw upon first-hand empirical research conducted in a range of geographical locations in Europe, North America, Australia, and China.

Such a comparative approach has at least two benefits to enhance our understanding of mega-city regions. First, it provides a series of situated accounts to inform specificity and varieties of the reconstituted state spaces, politics, and functionality around and across regions. Second, it can unravel generative conditions and circumstances through which new approaches in governance and planning are reshaping mega-city regions in divergent contexts. In this sense, the findings will offer an informed understanding if there are any common concerns and emerging trends underpinning these purported regional renaissances. To serve this purpose, the articles included in this volume address three interacting dimensions. These have an intrinsic value as a heuristic device with which to illuminate many other dimensions of city regionalism per se.

The first dimension relates to the rationales of governance and planning in mega-city regions. It invites us to uncover the indiscernible economic and political conditions that lie behind the emerging regional world. This further leads us to a deeper knowledge of the urban and regional regime constructed by various agents, including the state, when actively incorporating their functional importance into urban and regional governance. The second dimension is partly an institutionalized reflection of the regional representation of the state and other notable agents. This is expressed through the distribution of power among different agents, implementation of governance and planning policies, investment of major transportation and infrastructure, and institutional innovation. Of increasing importance is the structuration of states and how this contributes to our understanding of redefined state capacity in the design and delivery of urban and regional spatial strategies. The third dimension unpacks the consequences of mega-city regions' development. Are these regions imaginary spaces or new territorial structures with immense social, environmental, political, and spatial impact? How are they to be regulated? What are the future implications for them?

In what is to follow, we will provide a broad background about the transforming capital accumulation regime in relevant contexts that have strongly affected the formation of regions. This will be done by engaging with a circumspect assessment of inter-scalar politics in the process of reconstituted state spaces. Then we will directly confront the issues of governance and planning to examine how these practices might be conducted as a means of enhancing the power and/or regulation of key political agents in regional representation. A critical theme running through our discussion is the "politics of regional scale." This refers to the way regions are socially produced and politically transformed to become key arenas in and through which regulatory power is established.

Politics of scale and regional renaissances in advanced capitalism

The notion of "region" as a scale is in part a consequence of the collapse of the Fordist-Keynesian accumulation regime and the rise of the post-Fordism regime in advanced capitalism. A large body of scholarly literature has endeavored to capture this moment and traced the rise of more specialized systems of production and flexible accumulation in a growing, sophisticated global economy (e.g. Harvey 1990; MacLeod 2001; Brenner et al. 2003). Bob Jessop (2002) provides an insightful guide to identifying and understanding the profound changes in the territoriality of global capitalism associated with the restructuring of political and economic spaces. In retrospective discussions about the era of postwar Fordism, Jessop portrays a national space for economic organization and regulation. As summarized, the Keynesian welfare national state (KWNS) offered a distinctive set of national policies to support mass production and consumption from a demand perspective in capitalist cities. The justification of such an intervention on a national scale is market failure. Jessop (2002) further examines the reconstitution of the national territorial space when the capitalist state is transformed from KWNS to the post-Fordist accumulation regime, and to what he describes as the "Schumpeterian competition state." The new regulatory regime supports supply-side factors to develop the capacity of structural competitiveness and facilitate labor market flexibility and mobility. This defines a reworking of national territorial space, in which state functions are rearticulated upwards, downwards, and outwards so that place- and territory-specific strategies of economic development can be mobilized and achieved. To map this reterritorialization of modern capitalism, Scott (1998) demonstrates how such profound reshufflings give rise to a fourfold spatial hierarchy: the global scale, multinational blocs, sovereign states, and regions. The once single hegemonic national space has been reworked into deeply heterogeneous and contested spaces at the supranational and subnational levels (Swyngedouw 2000). In contrast to the Fordist era, we are witnessing a "relativization of scale" with no privileged level, yet assuming a preeminent role in the metagovernance of socio-economic affairs (Jessop 1999; MacLeod 2001: 11). This critical transformation of capitalism is closely intertwined with the successive rise of new territorial structures of political and economic activity. Various authors have sought to capture this moment. Tömmel (1997) proclaims the rise of multilevel governance in Europe. Swyngedouw (1997) enlightens us about "glocalization" and how the global, local, and other relevant geographical scale levels are the result and product of a deeply heterogeneous, conflictual, and contested process. One important consequence of this profoundly unfolding process of relativization of state space is that "region" becomes a focal point for economic governance.

Running parallel to this fundamental reconstitution of state territoriality is the rise of neoliberalism and the worldwide spreading of neoliberal economic and political policies in response to the crisis of the Fordist–Keynesian accumulation regime (Ma and Wu 2005). Market exchange is ethical in itself, and has become dominant in both thought and practice throughout much of the world since the early 1970s (e.g. Reaganism in the US). This powerful force of market revolution has resulted in immense multi-scalar deregulation, the removal of institutional constraints, expansion of the market power, more privatization, greater exploitation of labor, and significant liberalization of finance (Ma and Wu 2005: 3). One important consequence of this hegemonic discourse of market-led globalization is the emerging "new localism" of the 1980s to promote "zero-sum" politics of territorial competition (Peck and Tickell 1994) and a growing trend towards greater urban entrepreneurialism (Harvey 1989).

The restructuring of economic and political spaces and the spreading of neoliberalism are provoking some commentators to proclaim the hollowing out of and the imminent demise of the state (e.g. Ohmae 1995). When contemplating such a thesis, we might need to reflect on Jessop's (2002) two fundamental premises on the reconstituted role of the state. For Jessop, the capitalist mode of production is always an object of regulation by the state, because the market cannot be self-regulating. Such a regulating state "comprises an ensemble of socially embedded, socially regularized, and strategically selective institutions, organizations, social forces, and actions organized around (or at least involved in) the expanded reproduction of capital as a social relation" (Jessop 2002: 5).

Jessop's second premise is concerned with the process by which the capitalist state restructures itself in the face of regime shift, globalization, and regionalization. The overall trend is the establishment of new scales and arenas for state intervention. Brenner (1999: 447) argues at length that state reconstruction as a politics of scale can be construed as a sequence of grouping, trial-and-error strategies to manage the intensively conflictual forces of globalization through the continual construction, deconstruction, and reconstruction of relatively stabilized configurations of territorial organization. Jessop (2002) persuasively distinguishes three trends of state reorganization in capitalist transition, namely denationalization, destatization, and internationalization. He goes on to mention three counter-trends of state reorganization, namely a greater scope for the state in inter-scalar articulation, an increased role for the state in mega-governance, and interiorization of state reorganization in capitalist transition. Jessop indicates that these trends and counter-trends should not be deliberately presented in a one-sided and undialectical manner because they are intertwined with each other. Counter-trends must be viewed as "reactions to the new trends rather than as survivals of earlier patterns" in the face of globalization (Jessop 1999: 26). Crucially, the continuing strategic role of the national state should be stressed (Cameron and Palan 1999; Jessop 2002). The restructuring of the capitalist state should not be discussed in terms of the decline of the national state (Jessop 2002). A scaled state does not evolve into a zero-sum relationship. Rather, the state is reconstituted to solve the crisis and overcome the hurdles of capital accumulation and social reproduction in order to regain its functionality in glurbanization strategies.

Out of this innovative restructuring of political economic flows, new institutional spaces and new state spaces are being reforged with urban and regional scales coming to represent particularly significant strategic sites in the performance of accumulation, regulation, and political compromise (MacLeod 2001: 815). It is in this sense that the governance and planning of mega-city regions play a potent function in delivering the variety of regulatory spaces and needed facilities to lubricate capital flows. They also help to develop a context-specific synergy of collective actions to manage radical economic uncertainty in a fast-growing and fragmented political and economic space.

Politics of scale and regional renaissances in transitional society

Interestingly, this kind of creative restructuring of state territoriality is not limited to advanced capitalism. With the relentless pace of globalization, the developing world is experiencing a similar trend (e.g. Kelly 1997; Cartier 2005; Laquian 2005; Wu and Zhang 2007; Xu 2008). China is a prime example—its central state is confronting a series of immense challenges to its authoritative power and institutional capabilities because of decentralization and market reform. First, decentralization permits local states a wide range of economic responsibilities that used to be carried out by central government. Many socio-economic risks originally internalized and meditated on by the national state are now being externalized to local governments. Second, market revolution has had a tremendous effect on Chinese society. David Harvey (2005) argues that the spectacular emergence of China as a global economic power after the late 1970s was in part an unintended consequence of the neoliberal turn in advanced capitalism. China's neoliberal ascendancy seems to have developed in an unexpected direction. The result is that neoliberalism is growing rapidly in China, even as it festers and stagnates in capitalist economies.

In contemplating all this, Wu et al. (2007) provide an insightful interpretation on the rearticulation of state functions associated with the transformation of the regime of accumulation. Previously, socialist regimes in China were characterized by the extensive expansion of the means of production, constrained consumption, and forced organized labor process to achieve industrialization. The central state was the primary geographical scaffolding of accumulation and redistribution. In this sense, the central state hegemony was naturalized economically, with national space being the only legitimate object of economic management. However, the new regime after the economic reform features an intensified targeting of cities and regions as strategic sites for accumulation and regulation. As much as the pre-reform regime emphasized socialist industrialization through a redistributive function of the central state, the contemporary state regime involves a reconceptualization of cities and regions as the means to overcome the constraint of accumulation (Wu et al. 2007). It is in this sense that—perhaps more than in any other period over past decades—geographical scale in China has become a deeply diverse and contested terrain. Subnational units are prioritized scales for intensive accumulation, as seen through the increasing level of urbanization, the encouragement of rural–urban migration, the reinforcement of the central city's role in metropolitan governance, and the creation of competitive mega-city regions (see Xu 2008 for details). Urban and regional spaces are commodified through the land and housing markets. Foreign investment and joint ventures are being formed, adapting to the development of global capitalism. Many scholars have sought to capture this critical moment and to demonstrate how such profound transition reconstitutes territories into contested spaces and forces new actors (e.g. cities, regions, and non-state actors) into a more direct relationship with the outside world (Ma 2007; Wu et al. 2007; Xu 2008; Yeh and Xu 2008; Xu and Yeh 2009; Lin 2009). The former state system is penetrated by the market, by the international regimes, and by the growing discretion of local states. Indeed, Xu and Yeh (2009) raise comparative themes when discussing state reconstruction in contemporary China. They explore the present era to demonstrate a complex intertwining of institutions at all levels.

It is in this context that we witness a systematic reworking of state territorial space and function. The result is a burly wave of entrepreneurial mode of governance. Local states are preoccupied with offering supply-side policies to enhance place-specific socio-economic assets. Such features have provoked some commentators to declare the rise of Chinese local corporatist states (Oi 1995; Walder 1995; Duckett 1998), the demise of central government in local economic governance (Wong et al. 1995), the dehierarchization of socio-institutional spaces (Wu 2002), and the prominence of the market in establishing a full-fledged space commodification regime (Zhu 2002). Although some observers question the market applicability in China (Haila 2007; Holz 2007), most of them have developed theoretical discourses to suggest the powerful force of the economic process, i.e. marketization and decentralization of state functions to lower-level socio-economic institutions.

When reading these discourses, we have to be very cautious, because it must be stressed that any references to entrepreneurial governance, decentralization, and market reform are in no way indicative of the state's retreat. On the contrary, the remolded urban and regional fabric is increasingly underwritten by a more sophisticated state regime. Rather than retreating from urban and regional functions, the state needs to rebuild its capacity to insert a more qualitatively reorganized regulatory regime in local economic governance. Indeed, Wu (2002) argues that the total retreat of state power from economic and social life is too simplistic an interpretation of China's economic reform. The state apparatus has abandoned the direct allocation of production materials, capital, land, and, to a lesser extent, the workforce, while consolidating its regulatory power at the level of localities (Wu 2002: 1080).

In fact, this role that the state plays in economic governance is seen as critical in the context of a non-market society, for it helps and facilitates, among other things, the creation of market institutions, because the market cannot be perfectly self-regulating (Xu and Yeh 2009). This leads to the observation that, with a sophisticated governance structure, decentralization does not mean a reduction of state intervention and relaxation of control. Such claims become clearly evident when we inspect the changing role of the central state as a strategic site of socio-economic regime. As discussed, it is commonly believed that the dominance of the central state now declines in local economic development under conditions of preeminent decentralization and intensified globalization. Retreat from local governance was purposely designed by the central state during the early reform period (Xu and Yeh 2009). Qian and Weingast (1997) once observed that China's central government has explicitly limited its information as a way of credibly committing the center to avoid repeating the pernicious behavior of the pre-reform era. However, after reviewing recent development policies, it would be too simplistic to discuss state reconstruction only in terms of the decline of the central government and the rise of local authorities. Now, reasserting central state function is also purposely designed to regulate local economic governance, by reorganizing urban land governance, mediating through inter-scalar relations, controlling key strategic resources, and applying deliberate planning intervention in mega-city regions (Xu 2008; Xu and Yeh 2009). The central state continues to play a critical role in the ongoing transformation, even though the dominance at the national level is being decentralized.

Nevertheless, there is every reason to be profoundly concerned about the specificity of the state function in a transitional society. In such a context, the state governing society has a new meaning, because the state is more than just a market regulator; the state essentially becomes a market builder (Wu et al. 2007). In particular, state intervention carries a different meaning in the developmental state (Cumings 1999; Castells 2000), in which the state legitimizes itself by prioritizing development (Wu et al. 2007). Castells (2000) argues that a state is developmental when it establishes the principle of legitimacy and the ability to promote and sustain development. Drawing on the practice of Japanese industrial growth, the concept of the developmental state highlights the role of state bureaucratic organs in establishing strategic alliances with influential businesses to foster industrial expansion. This is similar to the experience of East Asian industrialization, where alternative governance was applied with the state as a machine governing the market (Wade 1990; Wu et al. 2007). The concept of the developmental state has immense appeal to scholars studying China, where state intervention is strong, but views are divergent as to what extent the Chinese government is developmental. Zhu (2004) describes a local state that is so developmental in nature that its pro-growth position undermines the capacity to exercise effective development control. Wu et al. (2007) argue that the Chinese state has moved away from being a resource distributor to being a market regulator, and is now finally an entrepreneurial market actor. Thus the corporatism that prevails in China is different from the developmental state, which is often presented at the national level and controls a set of policy parameters with greater national sovereignty.

Another key issue, in the present context, is how to look at the nature of state intervention. That the state continues to play a role in economic development at both central and local levels should not be read as the legacy of state socialism. Rather, there is a substantial recomposition in the nature of such interventions (Wu et al. 2007). Zhang (2000) assesses the transformation of the Chinese state from an anti-market totalitarian state to a largely pro-business authoritarian state. Particularly in developing urban and regional spaces, a number of scholars have also identified a qualitative new form of state that commodifies "place" as a space commodity (e.g. Yeh and Wu 1996; Yeh 2005; Wu et al. 2007). This consideration, together with those mentioned above, provides a state-theoretical perspective on which to interpret the reterritoriality of the state and the rise of regions.

It is in this context that we need to ask a complex question: how does regional development interact with different scales of regulation? Studying governance and planning in mega-city regions is the perfect medium to explore the intricate relations between regional resurgence and the changing nature of the state. Our major argument is that current regional governance and planning practices can be understood as important structural and strategic expressions of a locally, regionally, and globally articulated political process. They are being exercised not only as a new way to reposition regions in global capitalism, but also as a key political strategy of the states to reassert their functional importance in regional economic governance. In this regard, the point of Brenner et al. is well taken, as is their provocative presentation of the state–space relation—state spatial strategies must be viewed as historically specific practice through which state institutions attempt to adjust to the constantly changing geo-economic and geopolitical conditions in which they operate—their modalities, targets, and effects evolve qualitatively during the history of capitalist development (Brenner et al. 2003: 10). In the following section, we draw on the theoretical and practical development of regional governance and planning in different contexts to open up a venue for examining more closely the configuration of a new regime of regional political economic interests.

Mega-city region governance and planning

A message derived from preceding discussions is that the regional resurgence in both advanced capitalism and transitional economies (e.g. China) must be understood in direct relation to the contemporary process of state reterritorialization. Under this circumstance, regional scale has increasingly become a functional space for economic planning and political governance (Keating 1998). The pieces included in this volume were chosen to represent cases in three interesting contexts. First, we sought to include chapters that enlighten us on the European path to mega-city region governance and planning in a multi-level system. We use the appropriate phrase "multi-level system" (Marks et al. 1996; Blatter 2003) to present a depiction of contemporary reconstituted state spaces in European mega-city regions. In contrast to the state-centric thesis, such a system accentuates "the existence of overlapping competencies among multiple levels of governance and the interaction of political actors across these levels" (Marks et al. 1996: 41). Economic decisions and political influences over outcome are dispersed among contending subnational, national and supranational agents in rescaled state spaces. Despite this, city regionalism in Europe is often an institutionalized process in which supranational and national states can exert direct control over practices at the subnational level (see McCarthy 2000; Blatter 2003).

Second, we looked for cases that could provide insightful interpretations on the rise of mega-city regions in the US and Australia, two countries under federalism. Like European countries, both the US and Australia are currently experiencing a dramatic regional renaissance (Brenner 2002; Eversole and Martin 2005). However, even though numerous parallels exist between the European pattern and those in the US and Australia, the context of the latter is distinguished by a legacy of extreme jurisdictional fragmentation within their major city regions (Brenner 2002). Their histories of federalism result in extreme local control over economic development and a prevalent bottom-up approach in developing social and political institutions. Nevertheless, regional planning always involves some form of public intervention. Whatever the nature of that intervention, it will be interesting to witness the currently unfolding "relativization of scale" in the US and Australia in the face of the ominous economic downturn.

Third, we sought to include contributions that study mega-city regions in China, commonly known as a transitional society (Ma and Wu 2005). These contributions are quite meaningful, because the mega-city region as a novel spatial format was originally identified in areas such as the Pearl River Delta and Yangtze River Delta regions of China (Hall and Pain 2006). Our argument is that, regardless of whether China can be truly classified as a capitalist economy, Chinese cities and regions are being fundamentally reconfigured under the influence of powerful neoliberal ascendancy and rapid integration into the production of global commodities. It is this intrinsic link with global capitalism that works together with context-specific contingencies or path-dependencies in a post-socialist regime to provide a unique political and economic context. As such, China provides an excellent case for comparison.

Multi-level governance and planning in Europe

Throughout the history of advanced capitalism, the changing discourses of regional governance are often linked to the rearticulation of regional planning policies. Regional governance was first proposed to tackle the spatial mismatch between fragmented administrative boundaries and functional-economic territory (Brenner 2003). During the Fordist–Keynesian period, the main concern was on creating a territorial form of regulation to achieve administrative equalization and efficient delivery of public services. Since the early 1990s, the primary concern has centered on attaining economic priorities, such as territorial competitiveness and attracting external capital (Brenner 1999, 2003). New regional political-regulatory institutions are created to manage radical economic uncertainty in the post-Fordism era. Accompanying this regional renaissance is the rise of government-led strategic intervention. We identify three main typologies of regional institutions that underscore divergent planning philosophies and objectives.

At one end of the spectrum is the "reform-consolidation" concept, which favors political consolidation and strong institutionalization as the most effective means of achieving good governance (Bollens and Schmandt 1982; Lowery 2000). This approach occurred in the late nineteenth century and culminated around the 1950s and 1960s. Under the influence of Keynesian capitalism, state intervention was actively pursued to secure the general conditions required for the reproduction of both capital and labor through central state-led infrastructural investment and demand-side policies. Spatial production was organized primarily around the national territorial scale: local and regional states were understood as merely instruments of central state policies (Brenner 1997). The goal of state intervention was to establish a consolidated regulatory framework to guide outward urban expansion, to achieve planned decentralization and regional balance, and to reach efficiency in infrastructural provision. Spatial expansion was a strategy to achieve industrial growth. Planning was a blueprint to guide this expansion within a rationally and scientifically defined framework. Linking this type of comprehensive planning tradition with governance, Healey et al. (1997) uses the notion of the "provider" state to underpin the commanding actions of planning to control spatial organization and location of development on the national scale during the 1950s and 1960s.

At the opposite end is the "market public choice" concept, which looks at the individuality and multiplicity of fragmented regions as the most desirable way for better regulation (Tiebout 1956; Boyne 1996). One important consequence of this tradition is that neoliberal localism and urban entrepreneurialism prevailed during the 1980s. Many region-wide institutions were dismantled in Western Europe and North America. This heralded a retreat of state intervention from spatial formation, instead substituting a more deregulated approach to encourage the unfettered operation of the market. Spatial planning was thus in limbo, and perceived to exert a negative impact on wealth creation (Thornley 1993). An ad hoc and project-based planning approach was thus widely practiced to support private-sector development. This resulted in the planning system being left with little strategic capability.

However, these two extreme discourses in regional governance and planning have been under critical review. The consolidationist approach is too vulnerable to ideological attack, because it lacks political legitimacy and operates in an authoritarian manner (Lefévre 1998). The tradition of public choice is also subject to criticism, as it is deeply rooted in a neoliberal political environment, and can therefore produce external diseconomies (Briffault 2000). By the early 1990s, these debates had resulted in the evolution of a reactive interest in new forms of regional governance. "New regionalism" suggests a shift of institutional focus from the government to governance, and addresses an interactive approach through public-private partnerships, joint ventures, and cross-sectoral alliances (Jones 2001; MacLeod 2001). This discourse also stands to reason that the region is a significant and effective arena for situating the institutions of post-Fordist economic governance. The fascination with regional governance has been allied with widespread tryouts of new territorial forms, such as inter-government organizations, informal government partnerships, and functional consolidation (Rusk 1995). This political construction of institutional thickness prompts a systematic reworking of a hierarchical and functional planning towards a horizontal and network-based one (Williams 1999). Planning from this perspective means that cooperative thinking in bargaining arrangements and alliance building is valued (Healey et al. 1997). This has attributed to the revival of strategic planning that accentuates a more interactive process in a multi-level governance environment.

The contributions to Part I of this volume profoundly captures this key moment to demonstrate how innovation in governance and planning policies should face the complex socio-economic systems of European mega-city regions. In Chapter 2, Hall convincingly highlights how businesses interrelate and communicate in geographical space, with a particular focus on the emergence, dynamics, and planning of polycentric mega-city regions in contemporary Europe. Thus Hall explores the functional flows and radical reorganization of a fragmented landscape, in which mega-city regions serve as relatively fluid zones of transition between interpenetrating political and economic realms. As Hall's account clearly shows, the city region system is not constructed on a single scale. Rather, it grows out of a functionally networked but morphologically polycentric space that mediates flows at local, regional, national, and global scales. This, according to Hall, has immense policy ramifications and outcomes for infrastructure, transport systems, and regulation. "Muddling through the mucky field" demands an appropriate framework for "governance of flows" and "functional thinking" in spatial planning.

In Chapter 3, Salet raises analogous themes in his argument. He notes that it is these very inter-scalar and relational webs of multi-actor and multi-level governance that inspire planning innovation of local and regional public agencies. Salet develops this argument primarily through an interpretation of the shift of spatial form and governance structure in the urban network of the Randstad. On this basis, he elaborates the claim that novel regional planning strategies should be arranged through a completely new network of social interaction and practice to rectify the functional mismatch between the public sector-led planning and the private sector network. This seminal discussion of the Randstad cooperation illuminates the highly important directions for reforming state planning and institutions in what are increasingly multi-scalar and multi-centric political geographies.

Likewise, Albrechts critiques the stereotypical planning approach by proposing a new spatial planning strategy (see Chapter 4). For Albrechts, the new planning strategy is a transformative and integrative, public-sector-led, socio-spatial process through which visions/frames of reference, justification for coherent actions, and means for implementation are produced that shape and frame what a place is and what it might become. Albrechts' account further enlightens us about the construction of regional governance ensembles through the mobilization of a social base of support to resolve the conflicts between particular interests. This analysis, together with those by other contributors in

Part I, sketches an institutional approach that could prove most instructive in the search to comprehend the wider politically and socially constructed arena around which regions are configured, governed, and planned.

Multi-polity governance and planning in federacy

The notion of the "city region" is a new catchword in academic and policy literature in North America and Australia (Brenner 2002; Eversole and Martin 2005). However, this is not a modern innovation; rather, attempts at decentralizing governance to more local levels, including regions, have been part and parcel of the history of federalism in the two countries. There is a long list of planning and geography literature that is devoted to outlining the regional parameters of major urban problems, metropolitan regionalism, and the increasing interest in megaregions (Scott 1988; Soja 2000; Brenner 2003; Carbonell and Yaro 2005). In these imaginative and rousing academic inquiries, scholars portray the timeline of the recent rise of regional renaissance in North America and Australia. They observe numerous parallels in contrast to the European pattern in the evolutionary nature of regional planning and governance discourse (see last section). For instance, analogous trends are identified, permitting readers to comprehend how an interventionist, Keynesian, or high investment national planning approach in the Fordist regime undergoes a marked shift to a neoliberal approach in the post-Fordism era and then, more recently, to a more strategic and interactive model (Brenner 2003; Eversole and Martin 2005). No attempt is made here to repeat what we already know.

We stress differences. A federal structure, by nature, has the seeds of public policy fragmentation built into it. Blatter (2003) coins it a multi-polity system. In the American context, particularly following the imposition of Reagan's "new Federalism," the policy domain is fragmented vertically into state and local governments, and horizontally into special-purpose agencies and private capital (Brenner 2003). Moreover, America has a tradition of local control. In large metropolitan areas, the sheer number of local governments, each making decisions in their own self-interests, makes developing regional solution or regional institutions very difficult (Orfield and Luce Jr. 2009). This results in an inadequacy of regional planning capacity. For a long time, regional planning in the US has primarily been devoted to looking at the functional relationship between core cities and their surrounding small jurisdictions in a metropolitan context. There are no strategic interventions at the level of mega-city regions, barring a few historical exceptions such as the Tennessee Valley Authority in the 1930s (Dewar and Epstein 2007). Similarly, the fragmented and unstable nature of regional institutions in Australia makes it impossible to develop a strategically coherent framework for regions (Eversole and Martin 2005). There is little sign of practical movement by the state and federal agencies to coordinate their regional policies in a meaningful way due to the deficiency of inter-scalar flow. It is in this context that we find it fascinating to comprehend the rise of mega-city regions as particularly significant strategic sites in the performance of accumulation, regulation, and political compromise.

The contributions to Part II combine theoretical issues with more immediate questions of how regional geographies can be reworked in a period of rapid political and economic changes. In Chapter 5, Sassen provides a fascinating analysis of such a solution, with particular reference to the American case. She suggests that major shifts in the scales, spaces, and contents of economic activity are engendering novel spatial formats. Among the more prominent of these are global cities and megaregions, both of which are contributing to a whole series of old and new global intercity geographies. For Sassen, specifying a common analytic ground for these two very diverse spatial forms should enable us to develop a sharper approach to empirical research and, possibly, policy. As such, policy decisions should be made to maximize the combination of different locational logics by encouraging greater economic integration between a country's more globalized city (or cities) and its other areas currently performing subordinate functions within the national territorial hierarchy. In a more specific way, Sassen contends that mega-city regions can offer a variety of locations (nodes or spatially concentrated areas) that can suit different parts of a firm's supply chain. This is a point that echoes Soja's (2000) very innovative account on the deconstruction of the conceptual division between city and region. For Soja, it is time to shift the emphasis to the region, to absorb the urban into the regional, to see the urbanization process and the development of urbanism as a way of life and as simultaneously a process of regionalization and the production of regionality (Soja 2000: 179).

While Sassen focuses on the theoretical interpretation of city-region dynamics, the other contributors to Part II examine practical issues in regional governance and planning. For example, McCarthy (Chapter 7) addresses the institution and prospect for mega-city regions. By comparing four large urban agglomerations in both Europe and the US, she traces the growth of political fragmentation following the dissolution of the Fordist–Keynesian regime, and the rise of large metropolitan regionalism as a remedy. She explores the subtleties of cross-jurisdictional cooperation and competition in a multi-polity context. One of the key contributions of her chapter is to distinguish between the role of states in the European context and the American context: the former involves a stronger presence of national governments; the latter can be said to involve more powerful subnational states. McCarthy also underscores a disconnection between theory and practice. The notion of the "rescaling" of the national state whereby territorial competitiveness has increasingly come to be promoted at the scale of the city-region fails to operationalize the connections between large-scale restructuring on the one hand and metropolitan cooperation on the other. This implies that currently practiced regional projects are often contingent, politically charged and highly unstable institutional creations rather than obligatory and state-led responses to globally induced pressures or emergent governance problems.

In view of all this, Yaro elaborates a daring claim in his contribution to this volume (Chapter 6) that federal government should provide leadership in mega-city regions' development in the US. Yaro develops this argument primarily through interpretation of the enormous challenges now facing the US, including the growing concentration of population and employment in a handful of mega-city regions, intensifying global competition from Europe and Asia, climate change and the urgency to rebuild economic opportunities in underperforming communities and regions across the country. Yaro's decisive discussion implies that the current planning capacity in the US overprivileges state and local governments, as well as private investment, and thereby neglects the strategic priorities of the federal scale to guide another generation of growth that can be shared by every community and region across the country. Yaro's chapter thus usefully underscore the ways in which traditional federal countries see the promise of major policies and development initiatives finally moving ahead to herald a more strategic intervention at national and regional scales.

In Chapter 8, Abbott starts to uncover the workings of contemporary regional governance in Australia by exploring three leading regions of cities, namely Sydney, Melbourne and South East Queensland. He rediscovers the historical contingencies, governance and planning intensions for these areas, and explains how they have developed into mega-city regions in their own rights. Abbott argues that regional planning in today's Australia is dominated by the state government. Such practice faces dynamic conditions of complexity and uncertainty due to inadequate inter-scalar linkages. Strategic vision and planning capacity should be built up by an organized connectivity between key stakeholders. This implies a necessity for the Commonwealth and local governments to involve in regional planning governance and planning process.

Unlike Abbott, Harvey and Cheers focus on a rapidly deteriorating region by implicitly linking their discussion with wider political and social realities in Australia in Chapter 9. Australia's large geographic size and comparatively small population have meant that regions, in particular, are very meaningful (Eversole and Martin 2005: 5). However, the recent problems of regions have been paramount. First, administrative centers have often been geographically distant, and therefore out of touch with the needs of diverse local regions (Eversole and Martin 2005). This poses the difficulty of regulation at the regional scale, and originates a demand for localized decision making. Second, many regions have struggled to configure qualitatively new political and economic spaces in order to prevent their continuing erosion in the national and global economies. It is in this context that Harvey and Cheers investigate how a multi-centric city region in the Upper Spencer Gulf collectively resolves to reverse the decline. They implicitly suggest the defining feature of regions as historically embedded, socially constructed, and politically contested entities. They call for a multi-scalar and multi-agent approach to urban-regional structuring. The contributions included in Part II of this volume raise a number of fundamental questions about emergent forms of spatial organization in federalist countries where traditionally there has been little scope for central strategic intervention. Taken together, these works suggest that, while bottom-up initiatives continue to play key roles in regional structuring, there is a requirement for state strategies and state-led projects to be formed to bring some strategic coherence in pursuing the regional path to prosperity. The net outcome of this political structuration will reinforce the regional scale as important sites for accumulation and regulation.

State-led governance and planning under transition

Regional restructuring is a historically embedded process. There is no exception in transitional society, which carries strong legacies of its socialist history. Under state socialism, horizontal relationships among jurisdictions were not considered important, with hierarchical linkage dominating spatial formation (Xu 2008). This resulted in regions being dependent on the center. This dependency reduced regional policy to a sectoral policy (Gorzelak 1996) and within this context, socialist states often used frequent and arbitrary changes of territorial-administrative structure to serve the two primary purposes of clearing up the remnants of old regimes and enforcing central control (Solinger 1977; Surazska et al. 1997).

The transition towards a more market-oriented economy has created new conditions for regional development. However, the paths are now divergent to reflect context-specific governance problems. Central and Eastern Europe opted for neo-liberal shock therapy marketization. This was believed to be a quick way to close the wealth gap with the West and facilitate the process of returning to Europe. Ideological pressure for neoliberalism and "Europeanization" is so emotionally charged that the elites failed to consider alternatives and went for an overnight system-changing transition. As marketization proceeds at pace during the post-socialist period, creation of new regional institutions is widespread. In contrast to the West, regionalism in Central and Eastern Europe is not progressing out of fear of fragmentation or dysfunction of government services but with the objectives of realizing post-communist political imagination for decentralization, quick recovery of historical-cultural regional and local specificities, and Europeanization (Bialasiewicz 2002; Herd and Aldis 2003).

The objective of China's transition was not to propose the retreat of the state, which is different from the "shock therapy" in Central and Eastern Europe. Indeed, the success of China's gradual reform is often attributed to preserving state institutions while injecting market incentives (Wu et al. 2007). In conjunction with the introduction of a series of reform packages, there is a deep transformation to becoming a regime of accumulation that highlights cities and regions as strategic sites for regulation. A resultant new mosaic of regional development and territorial competition has begun to crystallize in China.

This crystallization adds to a heightened concern for growth, the maintaining of an efficient urban system, and farmland protection to herald an unprecedented revival of regional strategic planning in China's mega-city regions. However, the scale of urbanization has been vast and the speed of urbanization so rapid that it has threatened to overwhelm Chinese governments at various levels in managing urban and regional spaces.

The contributions in Part III try to connect post-reform state reterritorialization to this wider context and the rise of new socio-political space. This research direction examines how new state spatial strategies are getting reconstituted in a diffused power context and are engaging in new forms of struggle to redefine inherited forms of power in order to pursue new political agenda. In Chapter 10, for example, Xu and Yeh argue that regions can be best conceptualized as the always-contested spatial condensation for reconstructing state regulatory power. Using the Pearl River Delta's strategic planning as a case study, Xu and Yeh develop a state-theoretical interpretation of what is behind the increasing interest in this level of planning. They contemplate that current regional planning practice can be understood as an important political and strategic tool of state restructuration. Instead of treating regions simply as a spatial container for economic and social activities, regions themselves become critical elements as a medium for strategic intervention. This is more fundamentally related to the reconstituted state spaces in post-reform transition. In this sense, rather than shifting territorial development trajectories and coordinating regional growth patterns, regional strategic planning may appear to be little more than a cosmetic make-over that hides the intensifying interscalar competition within mega-city regions in China.

A number of closely related issues are raised in Gu and his colleagues' chapter on the spatial planning for urban agglomerations in the Yangtze River Delta (YRD) (Chapter 11). They open the chapter by outlining structural difficulties in forging coordinated growth in the YRD region. On this basis, they detail the spatial planning for the YRD as a state project. One important dimension of this chapter is the further depiction of the state as a powerful regional player in shaping territorial growth through large infrastructure projects and top-down state spatial regulation. This contribution, together with Xu and Yeh's, thus provide an initial set of conceptual tools through which to reinterpret the geographies of state space under transition. Both chapters denaturalize established assumptions associated with the decentralization of statehood and the resultant downward scalar shift of the state's function in capital accumulation and regulation. They explore the emergent character of state reconsolidation through state-led planning and its hidden and strategic agenda. This opens entirely new ways of looking at spatial planning as a tool to overturn established inter-scalar orders, in addition to its claimed rhetoric of sustainability and competitiveness. The contributions thereby illuminate many key aspects of reconstituted state spaces in a transitional society, as discussed in the preceding section.

Mega-city regions: convergence and divergence in governance and planning

The chapters in this volume have been chosen to address the theme of governance and planning of mega-city regions in Europe, America, Australia, and China. They develop arguments and interpretations under the conditions of intensified globalization and reconstituted state spaces. The picture of the mega-city region that they paint is unstable, fragmented, context-specific, contested, and politically charged. It does not easily lend itself to conceptualization or interpretation. Moreover, the contributions to this volume highlight the extent to which the enormous challenges have not yet been resolved through the construction of a new governance pattern and planning capacity. This should not come as a surprise if we reflect on some lively and provocative discussions in this volume. Contributors to Part I review and examine the Western European experiences, aiming to investigate how mega-city regions in a rescaling state context are reflected critically, creatively, and innovatively on the concepts and techniques used and the logic applied in tackling problems and challenges. It is discernible that there is a still a general lack of the right policy instruments to manage functional flow within mega-city regions. Within existing administrative structures, some policymakers think they have power, when in fact they are lacking it. Some have power but do not realize it, and thus there are both direct and indirect influences that can have unintentional consequences (see Chapter 2). Current European spatial policies may not be able to address these issues. This is both deeply problematic and quite inadequate if we are to fully develop the capacity of strategic planning.

Contributions in Part II analyze new approaches to address problems at the scale of the mega-city region in federacy. They provide an overview of a high degree of fragmentation and local control vis-à-vis the context, and ponder upon new city-region dynamics and rescaling planning capacity, considering inter-scalar and cross-sectoral governance as possible solutions. Previously, few believed that federal states needed regional strategies for promoting territorial growth and infrastructure investment. There are no longer any doubters. However, the problem is that both the US and Australia have yet to come up with institutional adequacy in both governance and planning to address either the explosive growth or dramatic decline of their megaregions in the global capital circuit. There is also much to be debated on how the mega-city region approach mediates between regional connectivity and political fragmentation, interdependence, and autonomy, and system-wide thinking and confinement to particular jurisdictions (Ross 2009).

In contrast to the context of Europe, the US, and Australia, China has witnessed the rise of regional strategic planning as a very powerful tool for spatial regulation. This is understandable in the sense that China has a strong state tradition and an enduring hierarchical state system. Nonetheless, the institutional capacity for strategic intervention is quite problematic because the function of regional planning is highly fragmented among different ministries and departments. Matters are further complicated by the top-down nature of strategic planning, which undoubtedly breeds tension, particularly with the hyper-competitive political environment tending to predominate, with local interests being undermined to various degrees, and with cities being accorded different bargaining power and political representation. Though scholars both inside and outside China advocate the governance approach and "learning from the West" to underpin the importance of an interactive and inter-scalar process in planning, the difficulty in its actual implementation is that it is very clearly related to the question of fundamental government reform and even political transformation in China where, traditionally, there has been a lack of an ordered and organized civil society (see Chapter 10).

It does appear, then, that we are witnessing in many countries the creation of heterogeneous regions in social, cultural, economic, and political terms. More interesting, perhaps, is the extent to which all of the contributors also accentuate the rise of regions as: (1) an emergent engine of national economic growth, which is closely tied in with the global capital circuit; (2) a venue of social and political transformations, such as dramatic demographic shifts, massive infrastructure investment, climate change, and environmental degradation; and (3) a new scale governing uncertainty and planning for prosperity, where we see the continued, if radically redefined, role of states in regulating inter-scalar relations and interceding with socio-political forces currently unfolding alongside globalization.

Drawing on these discussions, we can also identify some further questions that demand more systematic inquiry. Some selected questions are:

1. *In what sense are mega-city regions meaningful?* Paul Krugman's skepticism is perhaps useful to frame new research on the usefulness of a mega-city region. He wrote, "It's not at all clear to me that world competition is between mega-regions" (Krugman 2008). Much theoretical and practical work is still needed to explicate: what the mega-city regions can and cannot accomplish; why they are on such a different scale; and whether this scale can solve problems that cannot be achieved at other scales.
2. *In which way can the state deploy planning as spatial tactics* to regulate, produce and reproduce configuration of regional space for capital accumulation? Some chapters underscore the resurgence of regional planning in part as a state project. However, there remains a need to explore how the state power is forged into the regional matrix within which state intervention is to occur.
3. *What future directions can we identify?* The year 2008 has been a dramatic year for mega-city regions around the world. While the financial crisis has thrust us into a serious recession, we also see the promise of major policy and development initiatives—long advocated by regional scientists and planners—finally moving ahead in many countries. The chapters in this volume enlighten us that good governance and strategic planning are

unlikely to wane, albeit powerful neoliberalism's market revolution has persisted for decades in all contexts. Exploring further, the direction of governance and planning, both in theory and practice, is one of the most urgent intellectual and political tasks.

In sum, all the chapters that make up this volume provide an elucidation of mega-city regions—an emerging scale of capital accumulation and spatial regulation. They suggest that the governance and planning of mega-city regions have been uneven, taking different forms, in different places, and impelled by different context-specific forces and historical contingencies. They also indicate a common trend leading to the rise of mega-city regions as key sites, functions, and orchestrators of globalization, and the revival of regional planning as state intervention in a dynamically changing regional space. It is this distinctive combination of path dependencies and new regional initiatives that demand further research and more theorization.

Acknowledgment

The research for this chapter is supported by the General Research Fund of the Research Grants Council of Hong Kong (Project code: CUHK453609).

References

Bialasiewicz, L. (2002) "Upper Silesia: Rebirth of a Regional Identity in Poland," in J. Batt and K. Wolczuk (eds.) *Region, State, Identity in Central and Europe*, London: Frank Cass Publishers, pp. 111–32.

Blatter, J. (2003) "Debordering the World of States: Toward a Multi-level System in Europe and a Multi-polity System in North America? Insights from Border Regions," in N. Brenner, B. Jessop, M. Jones and G. MacLeod (eds.) *State/Space: A Reader*, Malden: Blackwell, pp. 185–207.

Bollens, J. and Schmandt, H. (1982) *The Metropolis: Its People, Politics, and Economic Life*, 4th edition, New York: Harper & Row.

Boyne, G. (1996) "Competition and Local Government: A Public Choice Perspective", *Urban Studies*, 33(4–5): 703–21.

Brenner, N. (1997) "State Territorial Restructuring and the Production of Spatial Scale: Urban and Regional Planning in the Federal Republic of Germany, 1960–90," *Political Geography*, 16(4): 273–306.

—— (1999) "Globalisation as Reterritorialisation: the Re-scaling of Urban Governance in the European Union," *Urban Studies*, 36(3): 431–51.

—— (2002) "Decoding the Newest 'Metropolitan Regionalism' in the USA: A Critical Overview," *Cities*, 19(1): 3–21.

—— (2003) "Metropolitan Institutional Reform and the Rescaling of State Space in Contemporary Western Europe," *European Urban and Regional Studies*, 10(4): 297–324.

Brenner, N., Jessop, B., Jones, M., and MacLeod, G. (2003) *State/Space: A Reader*, Malden: Blackwell.

Briffault, R. (2000) "Localism and Regionalism," *Buffalo Law Review*, 48(1): 1–30.

Cameron, A. and Palan, R. (1999) "The Imagined Economy: Mapping Transformations in the Contemporary State," *Millennium*, 28(2): 267–88.

Carbonell, A. and Yaro, R. (2005) "American Spatial Development and the New Megalopolis," *Land Lines*, 17(2): 1–4.

Cartier, C. (2005) "City-space: Scale Relations and China's Spatial Administrative Hierarchy," in L.J.C. Ma and F. Wu (eds.) *Restructuring the Chinese Cities*, Abingdon and New York: Routledge, pp. 21–38.

Castells, M. (2000) *The End of Millennium*, 2nd edition, Oxford: Blackwell Publishers.

Cumings, B. (1999) "Webs with no Spiders, Spiders with no Webs: the Genealogy of the Developmental State," in M. Woo-Cumings (ed.) *The Developmental State*, Ithaca, NY: Cornell University Press, pp. 61–92.

Dewar, M. and Epstein, D. (2007) "Planning for 'Megaregions' in the United States," *Journal of Planning Literature*, 22(2): 108–24.

Douglass, M. (2001) "Intercity Competition and the Question of Economic Resilience: Globalization and Crisis in Asia," in A. Scott (ed.) *Global City-Regions: Trends, Theory, Policy*, Oxford: Oxford University Press, pp. 236–62.

—— (2002) "From Global Intercity Competition to Cooperation for Livable Cities and Economic Resilience in Pacific Asia," *Environment and Urbanization* 14(1), 53–68.

Duckett, J. (1998) *The Entrepreneurial State in China: Real Estate and Commerce Departments in Reform Era Tianjin*, London: Routledge.

Eversole, R. and Martin, J. (2005) (eds.) *Participation and Governance in Regional Development: Global Trends in an Australian Context*, Aldershot: Ashgate.

Gorzelak, G. (1996) *The Regional Dimension of Transformation in Central Europe* London: Jessica Kingsley Publishers and Regional Studies Association.

Haila, A. (2007) "The Market as the New Emperor," *International Journal of Urban and Regional Research*, 31(1), 3–20.

Hall, P. (2001) "Global City Regions in the Twenty-first Century," in A.J. Scott (ed.) *Global City-Regions: Trends, Theory, Policy*, Oxford: Oxford University Press, pp. 59–77.

Hall, P. and Pain, K. (eds.) (2006) *The Polycentric Metropolis: Learning from Mega-City Regions in Europe*, London: Earthscan.

Harvey, D. (1989) "From Managerialism to Entrepreneurialism: the Transformation in Urban Governance in Late Capitalism," *Geografiska Annaler*, 71B(1): 3–17.

—— (1990) *The Condition of Postmodernity: An Enquiry into the Origin of Cultural Change*, Cambridge, MA and Oxford: Blackwell.

—— (2005) *A Brief History of Neoliberalism*, New York: Oxford University Press.

Healey, P., Khakee, A., Mottle, A., and Needham, B. (eds.) (1997) *Making Strategic Spatial Plans: Innovations in Europe*, London: UCL Press.

Herd, G.P. and Aldis, A. (2003) *Russian Regions and Regionalism*, London and New York: RoutledgeCurzon.

Herrschel, T. and Newman, P. (2002) *Governance of Europe's City Regions: Planning, Policy, and Politics*, London and New York: Routledge.

Holz, C. (2007) "Have China Scholars All Been Bought?" *Far Eastern Economic Review*, 170(3): 36–40.

Jessop, B. (1999) "Reflections on Globalization and its (Il)logics," in P. Dicken, P. Kelley, K. Olds, and H. Yeung (eds.) *Globalization and the Asia Pacific: Contested Territories*, London: Routledge, pp. 19–38.

—— (2002) *The Future of the Capitalist State*, Cambridge: Polity Press.

Jonas, A. and Ward, K. (2007) "Introduction to a Debate on City-regions: New Geographies of Governance, Democracy and Social Reproduction," *International Journal of Urban and Regional Research*, 31(1): 169–78.

Jones, M. (2001) "The Rise of the Regional State in Economic Governance: 'Partnerships for Prosperity' or New Scales of State Power?" *Environment and Planning A*, 33(7): 1185–212.

Keating, M. (1998) *The New Regionalism in Western Europe: Territorial Restructuring and Political Change*, Cheltenham: Edward Elgar.

Kelly, P. (1997) "Globalization, Power and the Politics of Scale in the Philippines," *Geoforum*, 28(2): 151–71.

Kidokoro, T., Harata, N., Subanu, L.P., Jessen, J., Motte, A., and Seltzer, E.P. (eds.) (2008) *Sustainable City Regions: Space, Place and Governance*, New York: Springer.

Krugman, P. (2008) *The Conscience of a Liberal: Mega-skepticism.* Online at http://krugman.blogs.nytimes.com/2008/04/12/mega-skepticism/ (accessed September 5, 2009).

Laquian, A. (2005) *Beyond Metropolis: The Planning and Governance of Asia's Mega-Urban Regions*, Washington, DC: Woodrow Wilson Center Press.

Lefévre, C. (1998) "Metropolitan Government and Governance in Western Countries: A Critical Review," *International Journal of Urban and Region Research* 22(1): 9–26.

Lin, G.C.S. (2009) "Scaling-up Regional Development in Globalizing China: Local Capital Accumulation, Land-centred Politics, and Reproduction of Space," *Regional Studies*, 43(3): 1360–591.

Lowery, D. (2000) "A Transactions Costs Model of Metropolitan Governance: Allocation versus Redistribution in Urban America," *Journal of Public Administration Research and Theory*, 10(1): 49–78.

Ma, L.J.C. (2007) "From China's Urban Social Space to Social and Environmental Justice," *Eurasian Geography and Economics*, 48(5): 555–66.

Ma, L.J.C. and Wu, F. (2005) "Restructuring the Chinese City: Diverse Processes and Reconstituted Spaces," in L.J.C. Ma and F. Wu (eds.) *Restructuring the Chinese Cities*, Abingdon and New York: Routledge, pp. 1–20.

MacLeod, G. (1999) "Place, Politics and 'Scale Dependence': Exploring the Structuration of Euro-Regionalism," *European Urban and Regional Studies*, 6(3): 31–53.

—— (2001) "The New Regionalism Reconsidered: Globalization, Regulation and the Recasting of Political Economic Space," *International Journal of Urban and Regional Research*, 25(4): 804–29.

Marks, G., Hooghe, L. and Blank, K. (1996) "European Integration from the 1980s: State-Centric v. Multilevel Governance," *Journal of Common Market Studies*, 34(3): 341–78.

McCarthy, L. (2000) *Competitive Regionalism: Beyond Individual Competition*, Washington, DC: US Economic Development Administration.

Ohmae, K. (1995) *The End of The Nation State*, London: Harper Collins.

—— (2001) "How to Invite Prosperity from the Global Economy into a Region," in A.J. Scott (ed.) *Global City-Regions: Trends, Theory, Policy*, Oxford: Oxford University Press, pp. 33–43.

Oi, J.C. (1995) "The Role of the Local State in China's Transitional Economy," *The China Quarterly*, 144 (December): 1132–49.

Orfield, M. and Luce Jr., T.F. (2009) "Governing American Metropolitan Areas: Spatial Policy and Regional Governance," in C.L. Ross (ed.) *Megaregions: Planning for Global Competitiveness*, Washington, DC: Island Press, pp. 250–79.

Peck, J. and Tickell, A. (1994) "Searching for a New Institutional Fix," in A. Amin (ed.) *Post-Fordism: A Reader*, Cambridge, MA: Blackwell, pp. 280–315.

Porter, M. (2001) "Regions and the New Economics of Competition," in A.J. Scott (ed.) *Global City-Regions: Trends, Theory, Policy*, Oxford: Oxford University Press, pp. 139–57.

Qian, Y. and Weingast, B.R. (1997) "Federalism as a Commitment to Preserving Market Incentives," *Journal of Economic Perspectives*, 11(4): 83–92.

Ross, C.L. (2009) (ed.) *Megaregions: Planning for Global Competitiveness*, Washington, DC: Island Press.

Rusk, D. (1995) *Cities Without Suburbs*, Washington, DC: Woodrow Wilson Center Press.

Salet, W., Thornley, A., and Kreukels, A. (eds.) (2003) *Metropolitan Governance and Spatial Planning: Comparative Case Studies of European City-Regions*, London and New York: Spon Press.

Sassen, S. (2001) "Global Cities and Global City-Regions," in A.J. Scott (ed.) *Global City-Regions: Trends, Theory, Policy*, Oxford: Oxford University Press, pp. 78–97.

Scott, A.J. (1988) "Flexible Production Systems and Regional Development: the Rise of New Industrial Spaces in North America and Western Europe," *International Journal of Urban and Regional Research*, 12(2): 171–85.

—— (1998) *Regions and the World Economy: The Coming Shape of Global Production, Competition, and Political Order*, Oxford: Oxford University Press.

—— (2001) (ed.) *Global City-Regions: Trends, Theory, Policy*, Oxford: Oxford University Press.

Scott, A.J. and Storper, M. (2003) "Regions, Globalization, Development," *Regional Studies*, 37(6–7): 579–93.

Scott, A.J., Agnew, J., Soja, E. and Storper, M. (2001) "Global City-Regions," in A.J. Scott (ed.) *Global City-Regions: Trends, Theory, Policy*, Oxford: Oxford University Press, pp. 11–32.

Simmonds, R. and Hack, G. (eds.) (2000) *Global City Regions: Their Emerging Forms*, London and New York: Spon Press.

Soja, E. (2000) *Postmetropolis: Critical Studies of Cities and Regions*, Oxford: Blackwell.

Solinger, R.J. (1977) *Regional Government and Political Integration in Southwest China, 1949–1954*, Berkeley, CA: University of California Press.

Stren, R. (2001) "Local Governance and Social Diversity in Developing World: New Challenges for Globalizing City Regions," in A. Scott (ed.), *Global City-Regions: Trends, Theory, Policy*, Oxford: Oxford University Press, pp. 193–213.

Surazska, W., Malikova, J. and Danek, P. (1997) "Towards Regional Government in Central Europe: Territorial Restructuring of Postcommunist Regimes," *Environment and Planning C: Government and Policy*, 15(4): 437–62.

Swyngedouw, E. (1997) "Neither Global nor Local: 'Glocalization' and the Politics of Scale," in K. Cox (ed.) *Spaces of Globalization*, New York: Guilford Press.

—— (2000) "Authoritarian Governance, Power, and the Politics of Rescaling," *Environment and Planning D: Society and Space*, 18(1): 63–76.

Thierstein, A. and Förster, A. (eds.) (2008) *The Image and the Region – Making Mega-City Regions Visible!*, Baden: Lars Muller Publisher.

Thornley, A. (1993) *Urban Planning under Thatcherism*, 2nd edition, London: Routledge.
Tiebout, C. (1956) "A Pure Theory of Local Expenditures," *Journal of Political Economy*, 64(5): 416–24.
Tömmel, I. (1997) "The EU and the Regions: Towards a Three-tier System or a New Mode of Regulation?" *Environment and Planning C: Government and Policy*, 15(4): 413–36.
United Nations (2004) *World Urbanization Prospect, Data Tables and Highlights*, New York: United Nations.
Vogel, R. (2010) "Governing Global City Regions in China and the West," *Progress in Planning*, special edition ("Governing Global City Regions in China and the West"), 73: 4–10.
Wade, R. (1990) *Governing the Market: Economic Theory and the Role of Government in East Asian Industrialization*, Princeton, NJ: Princeton University Press.
Walder, A.G. (1995) "Local Governments as Industrial Firms: an Organizational Analysis of China's Transitional Economy," *American Journal of Sociology*, 101(2): 263–301.
Williams, G. (1999) "Metropolitan Governance and Strategic Planning: A Review of Experience in Manchester, Melbourne and Toronto," *Progress in Planning*, 52(1): 1–100.
Wong, C.P.W., Heady, C., and Woo, W.T. (1995) *Fiscal Management and Economic Reform in the People's Republic of China*, Hong Kong: Oxford University Press.
Wu, F. (2002) "China's Changing Urban Governance in the Transition Towards a More Market-oriented Economy," *Urban Studies*, 39(7): 1071–93.
Wu, F. and Zhang, J. (2007) "Planning the Competitive City-region: the Emergence of Strategic Development Plan in China," *Urban Affairs Review*, 42(5): 714–40.
Wu, F., Xu, J. and Yeh, A.G.O. (2007) *Urban Development in Post-Reform China: State, Market, and Space*, Abingdon and New York: Routledge.
Xu, J. (2008) "Governing City Regions in China: Theoretical Discourses and Perspectives for Regional Strategic Planning," *Town Planning Review*, 79(2–3): 157–85.
Xu, J. and Yeh, A.G.O. (2009) "Decoding Urban Land Governance: State Reconstruction in Contemporary Chinese Cities," *Urban Studies*, 46(3): 559–81.
Yeh, A.G.O. (2005) "Dual Land Market and Internal Spatial Structure of Chinese Cities," in L.J.C. Ma and F. Wu (eds.) *Restructuring the Chinese City: Changing Economy, Society and Space*, London: Routledge, pp. 59–79.
Yeh, A.G.O. and Wu, F. (1996) "The New Land Development Process and Urban Development in Chinese Cities," *International Journal of Urban and Regional Research*, 20(2): 330–53.
Yeh, A.G.O. and Xu, J. (2008) "Regional Cooperation in the Pan-Pearl River Delta: A Formulaic Aspiration or a New Imagination?" *Built Environment*, 34(4): 408–26.
Zhang, T. (2000) "Land Market Forces and Government's Role in Sprawl – the Case of China," *Cities*, 17(2): 123–35.
Zhu, J.M. (2002) "Urban Development under Ambiguous Property Rights: a Case of China's Transition Economy," *International Journal of Urban and Regional Research*, 26(1): 41–57.
—— (2004) "Local Developmental State and Order in China's Urban Development during Transition," *International Journal of Urban and Regional Research*, 28(2): 424–47.

Part I

Multi-level governance and planning in Europe

2 The polycentric metropolis

A Western European perspective on mega-city regions

Peter Hall

European planners have recently become involved with two related concepts: the city region and polycentric development. This is reflected both at the European level, in the European Spatial Development Perspective (European Commission 1999) and the Leipzig Charter (EU 2007 DE 2007), as well as in European Union (EU) member states, such as the UK's Northern Way (Great Britain Office of the Deputy Prime Minister 2004). In addition, recent EU-funded research projects, such as ESPON and POLYNET, are centrally concerned with both concepts (ESPON 2006; Hall and Pain 2006). But in practice, each proves subtle and difficult to handle. City regions are *not* physical (morphological) units like the UK conurbations or the French agglomerations—nor are they administrative units, though administrative units must usually be used to define them. Essentially, they have to be defined in terms of Castells' "Space of Flows"—flows of people and information, especially the latter (Castells 1989: 344). They are Functional Urban Regions (FURs).

Recently, ESPON and POLYNET have independently extended the concept to a larger spatial scale: the mega, or mega-city region (MCR) (Hall and Pain 2006). This, of course, originated in the 1990s in Eastern Asia, where it was initially applied to areas such as the Pearl River Delta and Yangtze River Delta regions of China, the Tokaido (Tokyo-Osaka) corridor in Japan, and Greater Jakarta (Lin and Ma 1994; McGee 1995; Sit and Yang 1997; Hall 1999; Scott 2001; Lam 2002; quoted in UN-Habitat 2004: 63). It is essentially a new form—a series of anywhere between twenty and fifty FURs, each around a city or town, physically separate but functionally networked, clustered around one or more central cities, and drawing economic strength from a new functional division of labor. These places exist both as separate entities, in which most residents work locally and most workers are local residents, and also as parts of a wider functional urban region connected by dense flows of people and information along motorways, high-speed rail lines, and telecommunications cables. It is no exaggeration to say that this was the emerging urban form at the start of the twenty-first century. But its precise spatial form varies from country to country according to culture and planning regime: in the United States, lower density and less regulated, with "Edge Cities" or "New Downtowns" on greenfield sites, exclusively accessed by the private car; in Europe, medium density, regulated through green belts and other constraints, and centered on medium-sized country market towns or planned new towns (Garreau 1991; Scott 2001).

As basic building blocks, POLYNET used FURs, which it integrated into MCRs. (ESPON used similar concepts, but defined them in very different ways; Hall 2007.) POLYNET defined MCRs on the basis of contiguous FURs, thus producing eight MCRs (South East England; Central Belgium; Randstad Holland; RhineRuhr; Rhine-Main; Northern Switzerland; Paris Region; and Greater Dublin) in six EU15 countries, plus Switzerland, including a total of 135 FURs, and presented comparable data for three years (1980, 1990, and 2000, or national Census approximations thereto), though only five of its eight regions offer fully comparable data for population and employment over the two decades 1980–1990–2000.

POLYNET sought to define these MCRs through *analysis* of information flows in the knowledge economy—specifically, in the Advanced Producer Services (APS)—both through telecommunications and face-to-face, and both internal (node to node) and external. Its starting hypothesis was that these MCRs are becoming more polycentric—specifically, that MCRs are going through a process of "concentrated deconcentration" through a progressive redistribution of functions (Llewelyn-Davies 1997). In the FUR of the "first" city or cities, there is continuing concentration of higher-order service functions (financial and business services, design services, media, higher education, health); in FURs around secondary cities, there is growth of more routine functions (R&D, high-tech manufacturing; niche roles, such as university cities). However, all the cities and their functions remain highly symbiotic and abundantly interconnected. This networking through flows of information is a central feature and a vital key to understanding the phenomenon of the contemporary MCR.

POLYNET thus sought to address several research questions about the operation of MCRs. How are flows of information, through both physical face-to-face meetings and electronic communication, reconfiguring relationships within these regions? Are functional relationships between top-level and other centers changing? To what extent are these other urban centers dependent on, or independent of, service industry concentration in core cities? In what ways are these intra-regional changes affecting their relationships with other such regions, both within Europe and globally? In particular, to what degree are second-level centers in different regions communicating directly with each other, bypassing their respective core city regions? How far, and in what ways, are these patterns different, as between polycentric MCRs (such as the Randstad or RhineRuhr) and the much more "monocentric" Paris region? Can it be said that one pattern is more sustainable than the other, and if so, in what ways and to what degree?

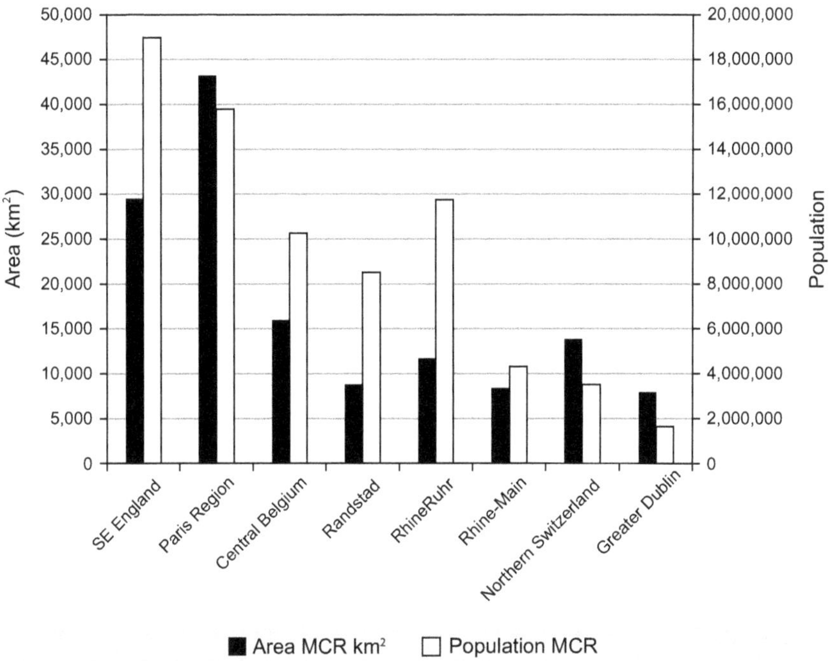

Figure 2.1 POLYNET MCRs: comparative areas and populations
Source: Hall and Pain 2006.

Defining the mega-city regions

Before addressing these questions, it was necessary to establish basic definitions and data. The eight North West European MCRs in the POLYNET study display a huge size range (see Figure 2.1). In area, they range from 7,800 square kilometers to 27,332 square kilometers (South East England) and 43,000 square kilometers (Paris Region); in population, from 1,600,000 (Dublin) to 15,600,000 (Paris Region) and 18,560,000 (South East England). But, interestingly, five out of the eight MCRs are within a much narrower range—between 8,000 square kilometers and 16,000 square kilometers in area, and between 8 million and 11 million in population.

They differ equally in terms of internal structure. Six MCRs are dominated by one central FUR: South East England by London (9.5 million), Central Belgium by Brussels (3.1 million), Rhine-Main by Frankfurt (2.4 million), Northern Switzerland by Zürich (1.1 million), the Paris Region by Paris (10.7 million), Greater Dublin by Dublin (1 million); two are strictly polycentric, with no dominant city: Randstad and RhineRuhr; but again, a majority have between five and twenty-five constituent FURs (see Figure 2.2). In terms of population

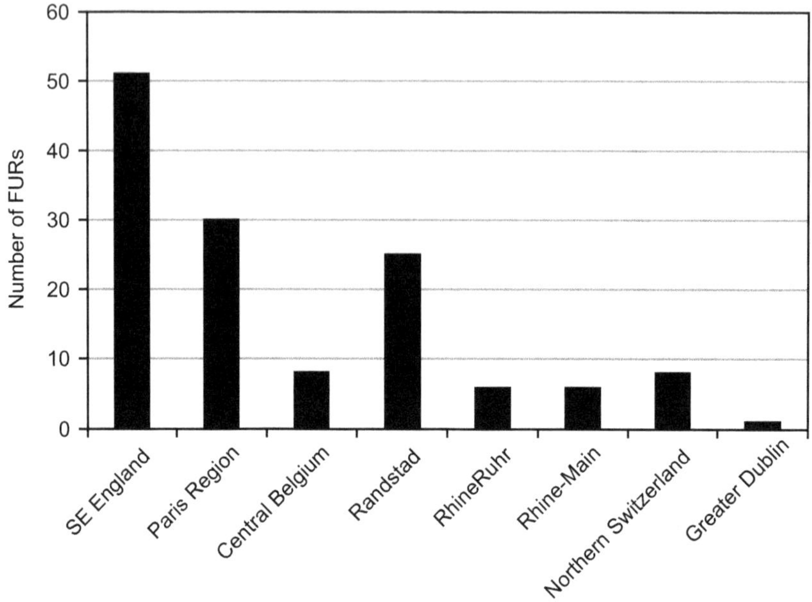

Figure 2.2 POLYNET MCRs: constituent FURS

Source: Hall and Pain 2006.

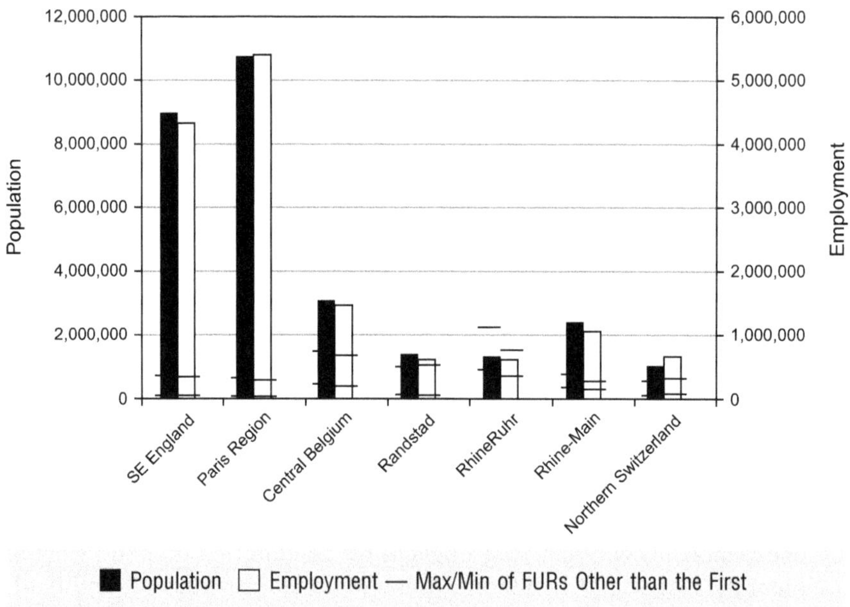

Figure 2.3 POLYNET MCRs: population and employment distribution in the first FUR, largest other FUR and smallest FUR

Source: Hall and Pain 2006.

and employment distribution, yet again the MCRs vary in polycentricity, but are predominantly "semi-polycentric" with one (or two) dominant FURs: Amsterdam-Rotterdam, Brussels, Frankfurt, Zürich, Paris (see Figure 2.3).

In their internal dynamics, the eight MCRs show a general trend towards decentralization. In South East England, the London FUR is growing, and the others even faster; in the Randstad, the four big city FURs are declining, and others are growing; in Central Belgium, growth is seen in Brussels and Flanders, but decline in the Walloon industrial belt; in RhineRuhr, core cities are losing to suburbs; smaller city FURs are growing. In Rhine-Main there is a complex picture: Frankfurt is growing (through migration from the five *Neue Länder*), while smaller FURs are growing faster, but Wiesbaden is losing. In the Paris Region, Paris and some distant FURs are growing, while the others are declining; there is a split between a dynamic west and a less dynamic east; in Dublin, the Dublin FUR is growing (but its growth is stagnating); smaller places are growing faster. For employment, the same trend to decentralization is evident (see Figure 2.4). In the Randstad, there is growth in (some) city

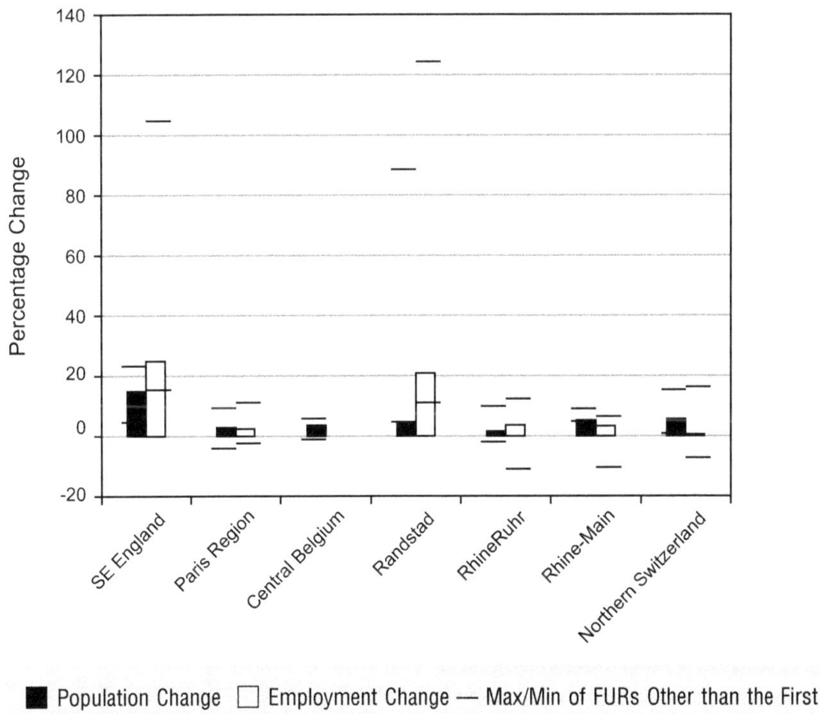

Figure 2.4 POLYNET MCRs: population and employment change 1980–2000 in the first FUR, largest other FUR and smallest FUR

Source: Hall and Pain 2006.

FURs, but faster growth outside; in Central Belgium, the Walloon industrial towns are losing; in RhineRuhr, the Ruhr FURs are losing, Rhine FURs gaining; in Rhine-Main, Frankfurt, and Mainz FURs are growing, but the rest are stagnating or losing; in Northern Switzerland, there is overall growth, but the FURs based on industrial towns have lost; in the Paris Region, Paris shows weak growth, some outer city FURs also have strong growth, but FURs around eastern towns demonstrate weaker growth.

In employment structure, there is a general shift from manufacturing and goods handling to services, especially Advanced Producer Services, but with considerable variations. Services are strong generally, especially in core city FURs, but are also growing at the periphery; the major exception is RhineRuhr.

Patterns of commuting are significant, because they give a first snapshot of internal flows of people within each MCR and thus a first clue to their degree of polycentricity. They show a general trend towards an increase in number and average length of commuter trips, with more trips "peripheral," i.e. not to the central "First City" FUR. But there are also some surprises. In the supposedly polycentric Randstad, there are no big flows from the south either into the Amsterdam core or between Randstad cities in general. In Central Belgium Brussels is dominant, but there are strong cross-flows and reverse flows between rings in the so-called Flemish Diamond north of Brussels. In RhineRuhr, Düsseldorf dominates the picture, but there are strong two-way-pair flows. In Rhine-Main, Frankfurt is dominant; here, peripheral cross-flows are negligible. In Northern Switzerland, likewise, Zürich dominates; here, there are some long-distance flows, as well as cross-flows between neighboring FURs. The Paris Region is highly monocentric, but there are some local commuter fields further out. Dublin displays extreme monocentricity.

Measuring polycentricity

The POLYNET study went on to pursue a key objective: to measure *the degree of polycentricity* in each FUR. Here, however, there is a basic distinction: between *geographical* or *morphological* polycentricity and *functional* polycentricity. Geographical or morphological polycentricity means that physically separate urban units, of similar size, co-exist in one region. Typically, however, ever since such polycentric regions as the Randstad and RhineRuhr were first identified in the 1960s (Hall 1966), the clear implication has been that they are characterized by a weak or flat hierarchy—there is no dominant city.

To measure geographical polycentricity, POLYNET employed three well-known and well-tested methods. The first, the rank-size rule, graphs the population size of each constituent FUR within an MCR on the y-axis against its population rank on the x-axis; the well-known assumption is that a slope of 45 degrees on double-logarithmic paper, a lognormal distribution, shows no tendency to primacy on the part of one or more dominant cities—the largest place would have a population double that of the next, and so on, without any sign of primacy at the top of the distribution. This analysis produces two distinct groups: a larger group of primate MCRs (South East England, Central Belgium, Rhine-Main, Northern Switzerland, Paris Region, Dublin), and a smaller group of non-primate MCRs (Randstad, RhineRuhr). But there was another significant finding: even in the first group, it appeared that there was, in effect, a semi-primate pattern, with one or more dominant cities superimposed on a predominant lognormal distribution. This was found not only in "primate" MCRs such as South East England and Paris but also in a "polycentric" region such as the Randstad; only RhineRuhr approximates to true polycentricity.

The second measure, self-containment, uses commuting data to generate an index and was first developed by Ray Thomas (1969). The key is the proportion of all workers who live and work in the same FUR within any given MCR—in effect, *lack* of commuting. It shows an interesting set of distinctions. In the Randstad the larger FURs are quite self-contained; in Central Belgium, all were self-contained; in RhineRuhr, they were quite self-contained, but to a declining degree; in Rhine-Main, the largest FURs were self-contained, but the others were not; in Northern Switzerland all were self-contained, but Zürich especially so; in the Paris Region, Paris was almost completely self-contained, but elsewhere there were outflows. However, beyond some critical commuting distance from the "First" City in each MCR—typically about 60 kilometers—FURs become quite strongly self-contained, with some 75 to 85 percent of working people living and working in the same FUR.

Finally, POLYNET developed entirely new measures of polycentricity (Green 2004). The index of Special Functional Polycentricity generated separate indices for in- and out-commuting between the constituent FURs within each MCR; the index of General Functional Polycentricity combined the two, and is the most useful general index of all. Each index can range from a theoretical minimum of 0.00 (no cross-commuting at all; complete self-containment) to 1.00 (complete exchange—everyone commutes from one FUR to another). The conclusion, at first surprising, is that in all eight MCRs there were very low degrees of polycentricity, ranging from only 0.02 in the Paris Region to a maximum of 0.15–0.20 in Randstad and RhineRuhr (see Table 2.1). But, as the Dutch team commented, this apparent paradox was only to be expected:

> A value of 1.0 would mean that all FURs in the greater Randstad area are equally well connected to each other in terms of commuter flows and that the entire working population works in a place different from their place of residence. Next to being a polycentric utopia, it would also be a clear recipe for traffic chaos and environmental degradation.
>
> (Van der Werff et al. 2005: 19)

Table 2.1 Special and general functional polycentricity indices of eight MCRs (2000)

MCR	*Special Functional Polycentricity Index (In-Commuting)*	*Special Functional Polycentricity Index (Out-Commuting)*	*General Functional Polycentricity Index*
South East England	0.14[a]	0.16[a]	0.15[a]
The Randstad	0.16	0.15	0.16
Central Belgium	0.04	0.04	0.04
RhineRuhr	0.20[b]	0.17[b]	0.19[b]
Rhine-Main	0.07	0.09	0.08
EMR Northern Switzerland	0.03	0.03	0.03
Paris Region	0.02	0.02	0.02
Greater Dublin	0.06	0.06	0.06

Notes:
a Calculation based on 2001 data with rings at NUTS (Nomenclature of Territorial Units for Statistics) 4.
b Calculation based on 151 NUTS 5 units, not on FURs.

Source: Hall and Pain 2006.

The problem of functional polycentricity: capturing flows of information

However, the real challenge for the POLYNET teams was to capture the second, *functional*, type of polycentricity through the measurement of flows of information, requiring different data and measurements (Hall and Pain 2006).

A first need is to understand how information moves for face-to-face communication. Over longer distances, it will continue to move by air, through the great international airports (Shin and Timberlake 2000). The main new influence is likely to be the development of the European high-speed train system, on present plans largely in place shortly before 2010 (Hall 1995). It is relatively easy to generate maps showing the connectivity of European cities in terms of the density of air and rail connections. But beyond this, measuring these "spaces of flows" within and between MCRs' flows proves difficult. A crucial first approach was a unique quantitative method developed by the GaWC Group (Taylor 2001). Originally devised to measure the "connectivity" between "global cities" arising from communications within worldwide APS

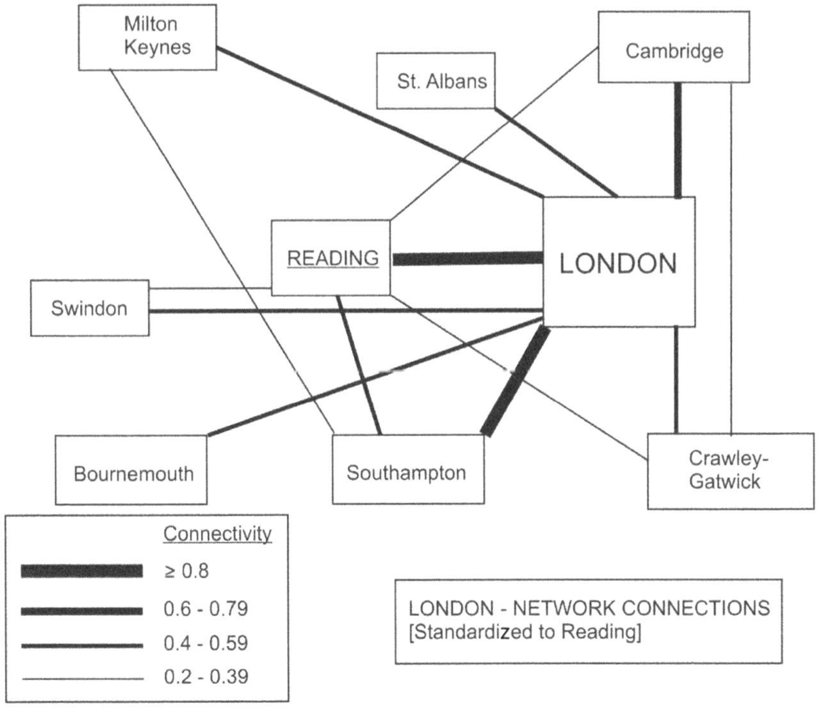

Figure 2.5 Intra-regional linkages for London

Source: Hall and Pain 2006.

networks, in POLYNET it was extended to measure the connectivity of *sub-global* European cities as well as global ones, by studying the importance of the business *functions* of APS offices located there. It further does so at four geographical scales: regional, national, European, and global (Taylor et al. 2006, 2007). Figure 2.5 is an illustration for South East England—it shows a degree of polycentricity, since some linkages bypass London, the "First City."

This entails analysis of very large primary data sets. In consequence, *transnational* inter-urban linkages between MCRs were not mapped in POLYNET, though data exist for a follow-up study. Also, to facilitate comparisons of polycentricity *between* MCRs, these inter-urban linkages are calculated for each MCR as a ratio between the connectivity of the most connected, or "First" city, and the second most-connected city for each region at the four geographical scales. The analysis showed that at the highest (global) level, the structure was much more primate than at regional or local levels. Within the RhineRuhr region, Cologne recorded 99 percent of Düsseldorf's local connectivity, but only 58 percent of its global connectivity. This, however, ignores the fact that there are significant differences in connectivity between these eight "First Cities." The global connectivity of London or Paris far exceeds that of Düsseldorf, yet this is not reflected in the comparative assessments. Finally, it needs emphasizing that this technique only provides a proxy for the strength and importance of the actual flows of information between the units within each MCR. How much information is actually passing between these cities—either virtually by email, by telephone, or inside people's heads through business travel—requires further analysis.

This was the focus of a second quantitative study, which focuses on capturing vital primary data on the actual flows of information passing within and between the MCRs. This proved a difficult matter. First, email traffic in particular is increasingly infected by spam and phishing. Even if spam filters are used to exclude this extraneous input, emails are increasingly used to convey relatively low-level routine information. It appears that non-work-related emails constitute some 40 percent of email traffic in the private sector and substantially higher in the public sector. Second, and more problematic, it has proved extremely difficult to obtain statistically significant samples of either e- or personal traffic. The POLYNET teams first hoped to analyze telephone traffic, with the assistance of telephone companies, and email data from sent-mail and in-boxes. But we were advised that this would contravene data protection legislation in EU member states, and we also discovered that there was no automatic or direct correlation between an IP address and the actual geographical location of offices. Finally, though French data have been successfully used to map telephone traffic in the Paris Region (Halbert 2004), such data are not generally available elsewhere.[1]

The POLYNET teams concluded that the practicable alternative was a web survey—we would ask respondents to keep a record of their telephone calls and email traffic and business travel for one week, and to report to us the top locations, discounting spam and other irrelevant mail. Then we approached Chief Executive Officers (CEOs), whom we were interviewing in another part of the study, asking them to cooperate in this work and to suggest names of colleagues. But despite an extension of time, the response proved generally poor. Out of the several thousand individuals that were approached, less than 10 percent (or 442 respondents) completed sufficient information for analysis. The Randstad and Paris Region teams achieved the best results, with over 100 results each: Northern Switzerland (over 70) and Central Belgium (over 50) were also satisfactory; RhineRuhr and Greater Dublin produced adequate returns (25–40 range), but Rhine-Main and South East England lagged with 20 results or fewer. For business travel, in particular, some results had to be discarded—some returns were left incomplete; others appeared implausible. Finally, there were no more than 46 complete records showing all kinds of contact for all the eight centers in aggregate; even then, some respondents failed to complete all the cells.

Even if the teams had been more successful, however, there would still have existed a major problem. Surveys of this kind can never convey the *quality* of the information exchanged. A single unit—whether a routine circular email or a message about a critically important business contract—is always given the same value. Respondents might have been asked to estimate importance on a simple linear scale, but given the poor response, that hardly seems plausible. The conclusion must be that quantitative measurement has limits, both in terms of practicability and the quality of output.

Capturing qualitative information

To tap into this qualitative evidence, POLYNET used semi-structured face-to-face interviews with CEOs and senior managers in the APS sectors. Developed from previous GaWC studies (Beaverstock et al. 2001; Taylor et al. 2003), they were designed to elicit "soft" evidence on the quality of flows—the production of knowledge, its transfer, and innovation—and on the interactions, not only between but also within cities. Hence, they demonstrated that purely quantitative measurement of information flows can never present an accurate picture of the volume and value of interactions, many of which are "invisible"—the most intense and important exchanges take place within globally networked First Cities. The most important interview findings (Hall and Pain 2006: Chapter 15) were:

- One city in each MCR constitutes the "First City" for global APS, with sectoral specialization in some cases. But secondary centers were also important: offices belonging to regional networks are spread out across secondary centers in each MCR, especially in accountancy; logistics has a distinct—and largely aspatial—servicing logic; banking/financial services, concentrated in First Cities, play an important role in fostering inter-relationships between sectors.

- The communication flows occurring within First Cities, and articulated through them, are of a far superior intensity and value to those occurring within firms across the wider MCRs. Locational concentration and clustering remain a key priority for most global firms across MCRs; there is no evidence that global functions are de-concentrating from POLYNET First Cities.
- There is limited evidence of functional linkages within MCRs. South East England is the major exception. Here, London's scale of global concentration masks significant functional connections not only between London and secondary centers but also between the secondary centers themselves.
- Paradoxically, morphological polycentricity may be associated with rather weak intra-regional functional linkages, as in the Randstad, while global concentration in London (treated as a monocentric area in the North West Metropolitan Area (NWMA) Spatial Vision) produces regional functional polycentricity.
- In all First Cities, the massive increase in email and use of intranet systems is not diminishing the absolute need for face-to-face contact, which is particularly associated with high-value exchanges. Most skilled front-office staff remains "locked into" clustered central city locations with an increasing need for travel, especially international travel. Car travel via motorway, as well as rail travel, is very important for intra-regional travel and access to airports for international travel, especially from First Cities. Also, e-infrastructure mirrors the patterns of intercity linkages, confirming the role of First Cities as "information gateways."
- First Cities have a unique regional role with respect to high-skill, specialized international labor supplies—competition for labor ties firms to specific central city locations, which depend on the residential preferences of employees.
- Locational decisions in APS are not based solely on rational economic criteria: an attractive "city environment" proves to be significant, but this is more about "city processes"—the "buzz" of the place—than physical infrastructure. Mobile high-quality labor is attracted to specific cities and places; office location and status are critical to the credibility of APS firms, and urban milieus are crucial for fostering innovation.
- While the scale of concentration differs from one First City to another, all have a distinctive "regional hub" function; in addition, London's global concentration gives it a unique role as a central "meeting place" in the APS "European region." Knowledge produced in—and dispersed through—international networks helps to build complementary functional relations, linking First Cities together.
- Geographical or administrative boundaries have little relevance to "natural" APS markets; national and international functions reside within a space defined by relations between major cities, while regional offices relate to local or sub-regional markets. Hence, the MCR is not something that respondents recognize—policy needs to take into account the varying

functions and linkages that underlie regional urban geography. Yet, they see that it has great policy importance in addressing areas that require non-market interventions—transport infrastructure, education, housing, and urban planning.

Implications for policy

Against the background of these findings, in the final research stage of POLYNET the eight teams turned to examine the policy context for their respective MCRs. They did so through two separate but closely linked exercises. First, they examined relevant policy statements—major EU, national, and regional policy documents—in order to establish how well they addressed the key issues identified in the research. Critical among these is the 1999 European Spatial Development Perspective (ESDP) and its subsequent influence on the 2007 Lisbon Treaty. Second, in a series of local focus group meetings they presented their findings to government and business stake-holders, and probed the views of their respondents on the key policy issues arising from the study and on the relevance of the empirical findings for planning practice, before meeting to compare and distil their findings.

The ESDP

The ESDP (European Commission 1999) has transformed *polycentrism* from a purely descriptive or analytic device into a policy instrument—a polycentric urban system is not merely something that one observes, for instance in Randstad Holland or in RhineRuhr, but a condition to which national spatial strategies should actively aim (Davoudi 2003: 991–95). The 2007 Leipzig Charter, likewise, commits planning ministers to "promote the establishment of balanced territorial organization based on a European polycentric urban structure." This is allied to concentrated deconcentration: a principle adopted long ago in Dutch spatial planning, which aims to disperse economic development from congested urban regions, but to reconcentrate it in urban centers in the less developed regions, thus benefiting both kinds of region.

The aim is not so much to redistribute some fixed amount of activity in a kind of zero-sum game; it is to encourage a significantly higher level of growth in less-developed regions and cities, some of which will be older industrial cities in need of restructuring, but a much larger number of which will be cities in the less densely-populated, less developed fringe regions of Western, Southern, Northern, and Eastern Europe. But here, it is necessary to recall one of the key conclusions above—polycentricity can apply at different spatial scales as well as in several dimensions. The ESDP does so at the largest possible geographical scale. Essentially, its main thrust is to divert some activities away from "global" cities such as London (and perhaps Paris) located in the so-called central "Pentagon" to "sub-global" or "gateway" national political or commercial capitals outside it (Helsinki, Stockholm, Copenhagen, Berlin,

Vienna, Rome, Madrid/Barcelona, Lisbon, and Dublin) and, since the EU's recent enlargement, the Eastern European capitals. They serve broad but sometimes thinly populated territories such as the Iberian Peninsula, Scandinavia, and East/Central Europe. Because they are national capitals serving distinct linguistic groups, they invariably have a level of service functions larger than would be expected on grounds of size alone; they tend to be national airport and rail hubs, and the main centers for cultural institutions and media on the national scale.

But here it is clear that polycentricity at one spatial scale may by no means be the same as polycentricity at another scale. In fact, by promoting polycentricity at a European or continental scale, it seems more than likely that the ESDP will generate monocentricity at a national scale—witness the concentration of activity and population in capital cities such as Lisbon, Madrid, Dublin, or Athens over the 1990s, or into Prague, Budapest, or Warsaw in the first decade of the twenty-first century.

At an even finer geographical scale, however, polycentricity can refer to the outward diffusion from either of these levels of city to smaller cities within their urban fields or spheres of influence. POLYNET shows that such a process has occurred on a wide scale around London, which is now the center of a system of some fifty centers within a 130-kilometer radius. Effectively, this implies the enlargement of the Pentagon itself, through simple outward expansion at its borders. London, Paris, Zürich, Frankfurt, Dortmund, and Amsterdam are all effectively on these borders; only Brussels is wholly within them; only Dublin is outside them. This is an issue in South East England, where UK government policy deliberately seeks to channel growth into broad growth corridors up to 90 miles (140 kilometers) north of London. It would promote greater polycentricity within an already polycentric region, but might detract from the goal of a more polycentric Europe—another paradox of scale.

At whatever scale, spatial planning strategies cannot impose rigid blueprints. They can only suggest broad desirable directions; since the ESDP is advisory, and the principle of subsidiarity will apply, implementation will come mainly at the national, regional, and local levels. Also, there can be no firm guarantee as to outcomes—increasingly, cities will compete directly in a global marketplace, and it can and should be no part of planning strategy to discourage this process. The European Union, however, will play an increasingly valuable role in coordinating efforts at these other levels, and in managing a variety of funds that can help shape them.

The notion of territorial cohesion, embodied in the EU's 1997 Amsterdam Treaty, is inherent in the ESDP. It seeks to promote spatial equity by maintaining key services even in remote areas. The concept stems from French planning traditions, specifically *aménagement du territoire*, reflecting the deep French influence inside DG XVI (DG Regio); this aims to achieve a harmonious allocation of economic activities. Territorial cohesion appears in the 2009 Lisbon Treaty on a par with economic and social cohesion, as a competence shared between the Commission and the Member States through the "community method" (Faludi 2003: 130–33; 2004a: 159–61; 2004b: 1019; 2004c: 1350–52; 2005: 1–3). There are unresolved issues here in relation to the 2000 Lisbon Agenda, which aimed to make the EU "the most dynamic and competitive knowledge-based economy in the world capable of sustainable economic growth with more and better jobs and greater social cohesion, and respect for the environment by 2010" (European Council 2000). Territorial cohesion might well suggest promoting growth in lagging regions, even at the expense of the most dynamic and competitive region in each country. In addition, as a decade of dynamic growth in the European economy gave way to recession in 2007–08, bringing economic contraction and rising unemployment, this was not a theoretical or academic issue. In the UK, for instance, studies suggested that the main impact of the recession had been felt not in the London economy with its dependence on financial services, as had been originally thought, but in regions such as the north of England, Wales and Scotland, where deindustrialization had weakened regional economies in the 1970s and 1980s. And this effect was likely to intensify as governments sought to cut public expenditure, because of the dependence of these regions on public service jobs (Sherwood 2009).

Key questions

POLYNET reached general policy conclusions through an intensive discussion process among the eight study teams. They are summarized below, as they appeared in the concluding chapter of the book (Hall and Pain 2006).

Key concepts—key myths?

Certain key concepts have come to dominate policy discussions in Europe, and often beyond its borders. Some, indeed, have come to assume an almost mythical or spiritual character, to be repeated as some kind of mantra without serious interrogation as to their meaning:

- *Polycentricity*: "Morphological polycentricity," which refers to the regional distribution of towns and cities of different sizes, is not the same as "functional polycentricity," which refers to flows of information and the organization of firms. A balanced spatial distribution of development does not guarantee an even distribution of complementary functions or a more sustainable form of development in the POLYNET MCRs.
- *Balanced development*: Indeed, the interview evidence paradoxically suggested that in reality, morphological polycentricity is associated with rather weak intra-regional functional linkages, just as, paradoxically, the depth of global concentration in London (regarded as a monocentric area in the North West Metropolitan Area (NWMA) Spatial Vision), was found to produce the most concrete evidence of regional functional polycentricity. Yet, this too is associated with uneven development, in the

form of an east–west economic imbalance in South East England. The growth of MCRs thus has implications for sustainable growth and social equity.

- *Sustainable development*: In spite of advances in information and communication technologies (ICT), face-to-face contact remains vital to the operation of advanced knowledge-based service functions. Polycentric regional development, whether functional or morphological, is also found to create "criss-cross" commuting that cannot be effectively supported by public transport. Functional polycentricity in APS is additionally associated with patterns of regional business travel that cross-cut the hub-and-spoke regional transport infrastructure. Hence both types of polycentricity have possible negative implications for environmental sustainability. How to overcome this—for instance, by developing more effective non-radial public transport links—represents a major challenge.
- *Economic competitiveness*: In all MCRs, functional concentration has been found to be essential to the development of international APS agglomeration economies and global business flows. First Cities—Dublin, London, Paris, Brussels, Amsterdam, Frankfurt, Düsseldorf and Zürich—play a vital "knowledge gateway" role, articulating their MCRs into the worldwide APS economy. Concentration of global functions and specialisms in these primary cities remains essential for high-complexity/high-value knowledge transfer, innovation, and production.
- *Spatial scale*: Polycentricity was found to be a scale-sensitive phenomenon. For example, Paris appears morphologically monocentric at the *regional* and *national* scales, but is functionally polycentric at the *global* scale. Functional polycentricity in APS at the *national* scale in Germany does not transfer across to the MCR *regional* scale—RhineRuhr and Rhine-Main have different spatial and functional configurations and connectivity to *global* APS networks. National contexts are important, but the MCR is identified as a vital spatial scale for spatial development policy and for the Lisbon Agenda of making Europe the world's "most competitive and dynamic knowledge-based economy" by 2010.
- *Sustainable management*: The MCR scale is hard to define, because APS flows are multi-scalar and do not coincide with administrative and political boundaries. Sustainable management of the MCR requires coordinated horizontal and vertical inter-organizational and cross-sector approaches. The concept of "economic competitiveness" should not be misconstrued as a "territorial competition" for inward investment. Cooperation is needed between cities and regions to reflect the functional complementarities that result from transnational knowledge-based networks; European policies can help to promote this.
- *Role of spatial planning*: The ESDP addresses problems of social and economic disparity by encouraging polycentric urban development to promote growth in less developed regions outside the "Pentagon." The

NWMA Spatial Vision aims to spread growth concentrated in London, Paris, and the rest of the "Pentagon" by improving accessibility and development of trans-European networks. POLYNET policy analyses for the eight MCRs suggest that spatial planning still places a heavy emphasis on physical infrastructure, but policymakers recognize the need for this to be complemented by new economic development approaches and "functional thinking."
- *Territorial cohesion*: The European Commission's concept of "territorial cohesion," developed in its Second and Third Reports on Economic and Social Cohesion (European Commission 2004) focuses on issues of "spatial equity" and uneven European spatial development, and stresses the need to promote social and economic balance across the EU territory. However, POLYNET findings on polycentricity suggest that balanced spatial development does not necessarily result in social equity and quality of life.
- *Spatial governance*: The globalization and liberalization of North West European markets for APS, strongly promoted by the European Commission, pose a challenge for the governance of flows which requires cross-jurisdictional and cross-sectoral structures. Over and above the specific spatial and functional features of the eight MCRs, the research has revealed important common policy dilemmas. In all cases, senior policymakers have described a serious lack of governance and policy instruments at the level of the MCR.

Sources of confusion

Arising out of this critical examination, we found many points where lack of clarity, as well as ambiguity, acted as barriers to coherent policy formulation:

- *Unclear interpretations of polycentricity*: The basic questions—"Is polycentricity more competitive?" and "Is it more sustainable?"—need to be addressed. The POLYNET results strongly suggest that these propositions are not simply self-evident; indeed, in certain cases, the opposite may be true. There is a basic need to distinguish between morphological polycentricity (generally interpreted in terms of commuting) and functional polycentricity (interpreted in terms of business travel and communication, especially face-to-face). The latter is more significant, and it is not axiomatic that it yields either more competitive or more sustainable outcomes than its opposite, monocentricity (or primacy). There is scope for a major new research program on this topic as an essential part of the revision of the ESDP.
- *Lack of understanding* of functional as opposed to morphological issues, especially in regard to polycentricity, leads to thinking solely in terms of physical infrastructure instead of "thinking functionally." Of course,

infrastructure is relevant to functional polycentricity. Accessibility within MCRs is important for face-to-face contact and vital for APS, so all regions need adequate investment in infrastructure and better management of transport networks to support major gateway cities and to strengthen external and internal flows. However, this is only part of the story.

- *The emphasis on self-containment and minimal travel* does not always accord with regional economic development priorities, particularly with the needs of knowledge-based business service suppliers and their workers; it therefore acts as an obstacle to regional economic development. Urban containment is assumed to be a realistic goal alongside polycentricity, but the reality is far more complex and even contradictory. Physical or morphological containment may, in fact, increase long-distance commuting, because it causes people to live farther away from their jobs than they would otherwise do.
- *There is a need for better understanding of the impact of functional polycentricity* on a number of key issues, including social equity and environmental sustainability. There are tensions, choices, and trade-offs here, as well as a basic policy paradox—that an even distribution of functions may conflict with environmental sustainability objectives. Equally, a failure to support regional development could inhibit economic growth. Thus, sustainable development may prove to be an oxymoron (Blowers and Pain 1999). The focus should be on addressing uneven economic and social equity as opposed to balancing the spatial distribution of development. Polycentric regional development does not necessarily enhance quality of life.
- *MCRs have common features, but are also diverse:* The ESDP therefore needs to be more scale-sensitive to Europe-wide, national, regional, and urban levels. There may be contradictions between policies at one scale and those at another. Thus, promoting polycentricity at the European scale (out of the central "Pentagon" to favor peripheral regions) may promote monocentricity at a national scale, as has been evident in Ireland and Southern Europe, and is becoming evident in Eastern Europe. Further, contrasts between national contexts may make one overriding European set of priorities inapplicable. The application of polycentricity to different spatial scales must be addressed, especially the national scale, not currently prioritized in the ESDP. This may have major implications for investment and the management of infrastructure.
- *There is also a need to understand functional specialization* across North West Europe's MCRs, in particular how this relates to the completion of the Single Market and the opening-up of the EU to global flows through World Trade Organization and G8 agreements. Morphological polycentricity appears to be associated with *sectoral* specialization between centers of a similar size, whereas a greater concentration of services and sectors in one city improves agglomeration economies and boosts regional functional polycentricity.

- *Contradictions between policy agendas*: There are implicit contradictions between the Lisbon Agenda and the ESDP on key issues, particularly on encouraging spatial competition versus spreading growth to lagging regions. *Changes to the ESDP are required* to incorporate Lisbon priorities for EU economic growth, regional equity, and sustainable development.

Implications for policy

- *Role of the ESDP*: In spite of the doubts expressed about its applicability to diverse national and regional circumstances, and possible contradictions with the Lisbon Agenda, the findings suggest that *the ESDP in fact requires more prominence* as a network for member states and regional cooperation. It currently has little visibility or recognition. It needs to be better understood because of its influence on national and regional planning in member states, and insofar as it is felt to have limitations and deficiencies, these need to be fully and expertly debated.
- *Policy coordination—how and where?* There is a diversity of policy frameworks and development perspectives for the MCR beneath the EU-wide level; but *is coordination either necessary or desirable at the level of the nation state?* When it is actually taken into account by policymakers, the ESDP is perceived as too abstract, dealing with a macro-level that does not relate to the diversity of circumstances at the MCR scale. Yet this scale has great importance for the Lisbon Agenda, and requires specific attention.
- *There is a need to involve the business community* to gain a better understanding of market drivers and conditions; inter-firm and inter-sectoral relationships; and economic and spatial relationships. Otherwise, policies may be based on a misunderstanding of the ways in which firms, especially but not exclusively in APS, may react. This is likely to represent a major challenge.
- *There is serious lack of governance almost everywhere at the scale of the MCR*, including a lack of policy instruments. Within existing administrative structures, some policymakers think they have power but, in fact, lack it. Some have power but do not realize it, and thus there are direct and indirect influences that can have unintentional consequences.
- *There is a need to think both strategically and locally.* To do this, it would be necessary to promote cooperative relations in MCR governance so as to reflect the network connections between cities across policy and sectoral fields at all geographical scales. There is likewise a need for cooperative MCR management networks—both vertical and horizontal.
- *Inter-regional competition for inward investment—and its converse in prosperous regions, local "Nimbyism"—need to be countered*, but this is hampered by the lack of governance structures for the MCR. Thus, there is a need to upgrade the entire MCR concept, including the injection of a democratic input at this scale.

- *The POLYNET MCRs are strongly interconnected through the global city network.* Support for transnational cooperation through INTERREG is necessary, because the MCRs are functionally interconnected and impact on each other. EU funding must continue to reflect this.

These questions provide a powerful agenda for further research.

Note

1 Follow-up research is beginning to analyze mobile telephone traffic for some European cities (Reades et al. 2007).

References

Beaverstock, J., Hoyler, M., Pain, K., and Taylor, P. (2001) *Comparing London and Frankfurt as World Cities: A Relational Study of Contemporary Urban Change*, London: Anglo-German Foundation.

Blowers, A. and Pain, K. (1999) "The Unsustainable City?" in S. Pile, C. Brook, and G. Mooney (eds.) *Unruly Cities? Order/Disorder*, London: Routledge, pp. 247–98.

Castells, M. (1989) *The Informational City: Information Technology, Economic Restructuring and the Urban-Regional Process*, Oxford: Basil Blackwell.

Davoudi, S. (2003) "Polycentricity in European Spatial Planning: From an Analytical Tool to a Normative Agenda," *European Planning Studies*, 11(8): 979–99.

ESPON (2006) *ESPON 1.1.1: Potentials for Polycentric Development in Europe*, Luxembourg: ESPON.

European Commission (1999) *ESDP—European Spatial Development Perspective: Towards Balanced and Sustainable Development of the Territory of the European Union*, Brussels: European Commission.

—— (2004) *A New Partnership for Cohesion Convergence Competitiveness Cooperation*, Luxembourg: Office for Official Publications of the European Communities.

European Council (2000) *2000 Lisbon Agenda*. Online at www.europarl.europa.eu/summits/lis1_en.htm#2 (accessed March 29, 2010).

EU 2007 DE (2007) *Leipzig Charter on Sustainable European Cities*. Online at www.eu2007.de/en/News/download_docs/Mai/0524-AN/075DokumentLeipzigCharta.pdf (accessed December 15, 2009).

Faludi, A. (2003) "Unfinished Business: European Spatial Planning in the 2000s," *Town Planning Review*, 74(1): 121–40.

—— (2004a) "Spatial Planning Traditions in Europe: Their Role in the ESDP process," *International Planning Studies*, 9(2): 155–72.

—— (2004b) "The Open Method of Co-ordination and 'Post-regulatory' Territorial Cohesion Policy', *European Planning Studies*, 12(7): 1019–33.

—— (2004c) "Territorial Cohesion: Old (French) Wine in New Bottles?" *Urban Studies*, 41(7): 1349–65.

—— (2005) 'Territorial Cohesion: An Unidentified Political Objective: Introduction to the Special Issue, *Town Planning Review*, 76(1): 1–13.

Garreau, J. (1991) *Edge City: Life on the New Frontier*, New York: Doubleday.

Great Britain Office of the Deputy Prime Minister (2004) *Making it Happen: The Northern Way*, London: ODPM.

Green, N. (2004) "General Functional Polycentricity: A Definition," *POLYNET Working Papers*, London: Institute of Community Studies/The Young Foundation & Polynet Partners.

Halbert, L. (2004) *Densité, Deserrement, Polycentrisme et Transformation Economique des Aires Métropolitaines (Density, Deconcentration, Polycentricity and Economic Transformation of Metropolitan Areas)*. Thesis submitted for the Degree of Doctor in Geography of the University of Paris-I (Sorbonne) (in French).

Hall, P. (1966) *The World Cities*, London: Weidenfeld & Nicolson.

—— (1995) "Towards a General Urban Theory," in J. Brotchie, M. Batty, E. Blakely, P. Hall, and P. Newton (eds.) *Cities in Competition: Productive and Sustainable Cities for the 21st Century*, Melbourne: Longman Australia, pp. 3–31.

—— (1999) "Planning for the Mega-City: A New Eastern Asian Urban Form?" in J. Brotchie, P. Newton, P. Hall, and J. Dickey (eds.) *East West Perspectives on 21st Century Urban Development: Sustainable Eastern and Western Cities in the New Millennium*, Aldershot: Ashgate, pp. 3–36.

—— (2007) "Measuring a Polycentric Europe: Whose Space of Flows?" in Various Authors, *Geophilia: O Sentir e os Sentidos da Geografia: Homenagem a Jorge Gaspar* (*Geophila: The Experience and the Meanings of Geography: Homage to Jorge Gaspar*), Lisbon: Centro de Estudos Geográficos da Universidade de Lisboa, pp. 613–35 (in Portuguese).

Hall, P. and Pain, K. (2006) *The Polycentric Metropolis: Learning from Mega-City Regions in Europe*, London: Earthscan.

Lam, W. (2002) *Race to Become China's Economic Powerhouse*, CNN, June 11. Online at http://edition.cnn.com/2002/WORLD/asiapcf/east/06/11/willy.column/ (accessed on March 29, 2010).

Lin, G. and Ma, L. (1994) "The Role of Towns in Chinese Regional Development – The Case of Guangdong Province," *International Regional Science Review*, 17(1): 75–97.

Llewelyn-Davies (1997) *Four World Cities: A Comparative Study of London, Paris, New York and Tokyo*, London: Comedia.

McGee, T. (1995) "Metrofitting the Emerging Mega-Urban Regions of ASEAN: An Overview," in T. McGee and I.M. Robinson (eds.) *The Mega-Urban Regions of Southeast Asia*, Vancouver: University of British Columbia Press, pp. 3–26.

Reades, J., Calabrese, F., Ratti, C., and Sevtsuk, A. (2007) "Cellular Census: Explorations in Urban Data Collection," *Pervasive Computing*, 6(3): 30–38.

Scott, A. (ed.) (2001) *Global City-Regions: Trends, Theory, Policy*, Oxford: Oxford University Press.

Sherwood, M. (2009) "Recession Creates a New Twist on Old North/South Divide Story," in *The Insight Report: An Experian Report*, 2: 13–19. Online at www.experian.co.uk/www/pages/downloads/insight_report/experian_q2_insight_report.pdf (accessed December 15, 2009).

Shin, K.H. and Timberlake, M. (2000) "World Cities in Asia: Cliques, Centrality and Connectedness," *Urban Studies*, 37(12): 2257–85.

Sit, V. and Yang, C. (1997) "Foreign-Investment-Induced Exo-Urbanisation in the Pearl River Delta, China," *Urban Studies*, 34(4): 647–77.

Taylor, P. (2001) "Specification of the World City Network," *Geographical Analysis*, 33(2): 181–94.

Taylor, P., Beaverstock, J., Cook, G., Pandit, N., Pain, K., and Greenwood, H. (2003) *Financial Services Clustering and Its Significance for London*, London: Corporation of London.

Taylor, P., Evans, D., and Pain, K. (2006) "Organisation of the Polycentric Metropolis: Corporate Structures and Networks," in P. Hall and K. Pain (eds.) *The Polycentric Metropolis: Learning From Mega City-Regions in Europe*, London: James & James/ Earthscan, pp. 53–64.

—— (2007) "Application of the Inter-Locking Network Model to Mega-City Regions: Measuring Polycentricity Within and Beyond City-Regions," *Regional Studies* 42(8): 1079–93.

Thomas, R. (1969) *London's New Towns: A Study of Self-contained and Balanced Communities*, London: PEP.

UN-Habitat (2004) *The State of the World's Cities 2004/2005: Globalization and Urban Culture*, London and Sterling, VA: Earthscan.

Van der Werff, M., Kapoen, L., Kloosterman, R. and Lambregts, B. (2005) *POLYNET Action 1.1: Commuting & the Definition of Functional Urban Regions: The Randstad*, London: Institute of Community Studies/The Young Foundation and Polynet Partners. Online at www.polynet.org.uk/ (accessed December 15, 2009).

3 Innovations in governance and planning

Randstad cooperation

Willem Salet

Introduction: different styles of urbanization

This chapter deals with the innovation of governance and planning in European regions in general, and more specifically, the urbanized Randstad area of the Netherlands. As one of the smallest countries of Europe, Dutch society has always been intensely interconnected with the outside world. The Dutch economy serves as the gateway to Europe, with its strength in international trade and services. Education and culture are well linked with the outside world, albeit with different accents during the last century. The prewar generation was strongly influenced by dominant European countries, namely the UK, Germany, and France. My own generation grew up in the postwar circumstances of the 1950s and 1960s. As with everywhere else in Western Europe, this generation became overwhelmingly dominated by North American influences. Even in the critical students' epoch of the early 1970s, I estimate that about 80 percent of our academic textbooks came from the US. The current generation of students is no longer solely influenced by North American culture, and the gradual unification of Europe has generated a new fascination for countries within Europe itself—particularly, this includes many of the new member states in Central and Eastern Europe—as well as the booming countries in Asia. We are well aware that history is changing at a faster rate and on a larger scale than ever before. With more economic and cultural potential in the "new world" than ever expected, countries such as China and India, as well as the region of East Asia, are setting new trends, not just in economic development but in many facets of explosive urbanization. It is in these emerging climates that Western engineers, urban planners, and architects have the opportunity not only to apply the best of their capabilities but also, increasingly, to learn about changing urban forms and innovative ways to deal with these explosive conditions.

In their turn economic specialists and urban experts in China are very keen to learn from Western experiences. As a result, one may observe a growing international exchange of knowledge, with a focus on the economic and urban development of Asia. Considering the number of students abroad, the younger generation of Chinese students appears to be particularly attracted to the knowledge and prestige of North American schools. There may be an economic rationale behind this "North American gaze," but urban development is not just a matter of economic growth. Urban growth is an integral "oeuvre," as the famous French urban sociologist Henri Lefebvre used to describe the myriad of social, cultural, ecological, and economic experiments of urban evolution (Lefebvre 1996). The French word "oeuvre" cannot easily be translated into English and certainly not in the way Lefebvre used it. He used the term to indicate the ongoing evolution of cities under construction—not deliberately planned out of one integral mindset or of one planning authority, but building forward on a plethora of past experiences and made up as a sort of urban laboratory in a continuous struggle between completely different spheres of urban activity (economic, social, cultural, etc.). Obviously, North American cities have their own history and oeuvre as well. It is not prudent to qualify these cities in a much generalized sense, but there are some basic factors that underlie local differentiation in both continents. In Europe, cities usually crystallized in completely different, and more variegated, social and economic forms, compared to the prototype of the property-led North American city. History matters in European cities, in particular on the continent, where the bourgeois kept hold of inner cities, even in the industrializing mid-nineteenth century. In Europe, the historical center has been the center of urban development for centuries, forging more city-centered paths of urban development than the outward urban development in the US (Fishman 1987). Another basic difference is that in Europe, the rigid selections of the market mechanism have traditionally been mitigated by deliberate policies of social development and urban containment more than in the liberal North American context (Hall et al. 1973; Marcuse and Van Kempen 2000). European cities build forward on urban heritage, provide social amenities such as universal education and health, and offer extensive quarters for social housing and public sector financed systems of public infrastructure. Care for the environment is institutionalized in urban and regional policies, too. These characteristics are not typical in the liberal American society, where market-led urbanization is the standard.

In this chapter, we will focus on the challenge to create integrated "urban use value" rather than just "commercial value" in processes of urbanization (Lefebvre 1991). This requires the integration of different economic, social, and ecological qualities. In doing so, it has to be recognized that cities are in various stages of rapid transformation (this also applies to European cities). The compactness of urban expansion and the restrictive physical containment of dynamic urban development are in contrast to the wideness of green surroundings that, up until the last two decades, were predominant in almost the entire length of the previous century. Although metropolitan regions tend to grow, urban activities are getting more and more decentralized within these wide regions, developing into new patterns of polycentricism and mixed varieties of citifying landscapes (for the contrasts over thirty years, see the publication by Hall et al. on urban containment in 1973, and Hall's recent book on polycentric regions in 2006, which is highlighted extensively in Chapter 2 of this volume; Hall et al. 1973; Hall and Pain 2006). On the one hand, processes of worldwide economic liberalization and economic globalization are occurring, while on the other hand, continuing individualizations of the use of spaces are in effect. Together, these are now the driving forces of new urbanization processes. The familiar balance of urban compactness has been disturbed. In addition, the context of governance has strongly changed in multi-actor and multi-level arenas. The challenge for spatial planning strategies is to find new answers to guide and integrate the economic, social, and ecological processes in this dynamic context in new and balanced ways. We will discuss the strategies that have been developed in the Randstad in the Netherlands. Before focusing on the Dutch planning strategies we will analyze in a more general sense the current changes in cities in Europe, the driving forces of these changes, and the new conditions of metropolitan governance.

The current changes in cities in Europe and their driving forces

The rescaling of economic and social parameters since the early 1990s has had a dramatic impact on the spatial configuration of city regions in Europe. The current transformation of cities is not just a matter of urban expansion. Urban expansion is clearly an important and almost permanent drive for the enlargement of scale in many urban regions in Europe throughout the whole of the last century. This certainly goes for the regions in the most urbanized zone of Europe, which meanders from the Midlands in England, through London and South East England, and through the intermediary zones of Paris, Brussels, and the Dutch Randstad, via the German Ruhr Area and South East Germany to northern Italy. A secondary zone of trans-European urbanization runs along the Mediterranean Sea (from Valencia and Barcelona in Spain, through southern France to the Italian coast near Genoa). Beyond these large-scale European zones of urbanization, one may also find shrinking cities in some peripheral regions (for instance, in parts of eastern Germany and Poland).

Longitudinal investigation confirms the continual growth of these particular metropolitan regions at the expense of the non-metropolitan regions in Europe (ESPON 2006) (see Figure 3.1). The growth of urban regions reflects the expansion of the population, the development of urban activities (in particular of the labor markets), and the increasing consumption of space for urban activities. However, the ongoing enlargement of urban scale may explain only part of the changing landscapes of the European cities. Even more dramatic are the growing scope of urban activities and the underlying dynamics of rescaling social and economic activities, criss-crossing through the territorial coherences of urban and regional communities. The familiar territorial

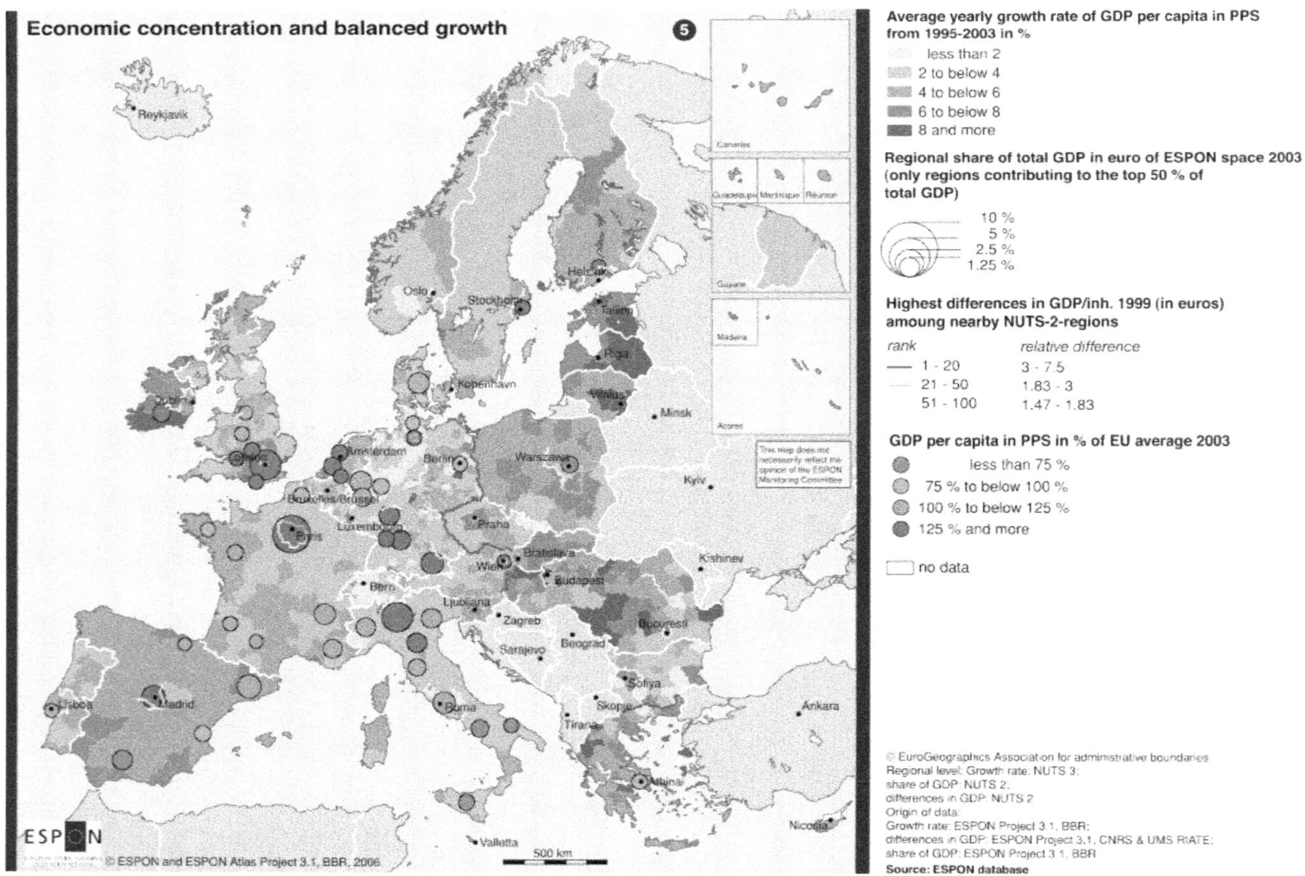

Figure 3.1 Economic concentration and balanced growth in Europe
Source: ESPON 2006.

coherences of centripetal cities are being disrupted as a result of this dynamic rescaling.

We will discuss two sides of the rescaling processes. The first concerns the *macro-sociological and economic trends* of globalization and modernization that generate cities with increasingly trans-national economic relationships and trans-national cultures and urban populations. Enhanced by new ways of telecommunication, economic and social activities have become far less bounded by place. The moving abroad of locally based activities is not a new phenomenon—urban activities have always had an external horizon. What is really new (certainly in its current grade of intensity), is that the nature of relationships has radically evolved into trans-national forms. An example: the seaport of Rotterdam was made the largest seaport in the world in the 1960s through the joint efforts of the government and the proud, local entrepreneurs of Rotterdam. Forty years later, Rotterdam is still a major international seaport, but now it is hardly possible to find an indigenous entrepreneur affiliated with the urban community of Rotterdam. The globalization of economic relationships, the continual specialization of production and consumption, and the differentiation of labor are the economic drivers of change that result in new patterns of functional spatial coherence. The places of connectivity are becoming more important. As Melvin Webber had predicted in the early 1960s in a forward-looking essay, non-place realms have become at least as important as locally bounded characteristics (Webber 1964). As a consequence, proximity is no longer the crucial dimension of urbanity, but connectivity. Being connected with the outside world in different social and functional networks has become a decisive criterion for economic and social evolution. Webber rightly concluded that density of interaction does not always require physical proximity; social interaction may have a high thickness in extensive, yet physically thin urban areas as well (Webber 1964). This focus on functional rather than territorially bounded spaces is not to claim that local or regional place no longer matter. Characteristics of place are still very important conditions that make some places more or less attractive for social and economic activities. Place certainly matters in a world of increasingly non-place bounded interrelationships, but not automatically in a centripetal sense. Furthermore, persons and organizations act in different places simultaneously. Multiplicity of urban activities has become a regular dimension of the new urban reality; it has brought non-place networks of interdependency, the juxtaposition of completely different experiences within territorial spaces, and new sorts of exchanges within territorial places (Amin and Thrift 2002).

The macroeconomic interrelationships of activities at different spaces have been intensively investigated since Sassen instigated the line of inquiry into new spatial hierarchies in global financial and economic networks (Sassen 2001; Swyngedouw and Baeten 2001; Taylor 2003; Brenner 2004; also see the contribution by Saskia Sassen in this volume).

Rescaling of the urban experience is also occurring at the *micro-level of urban lifestyles and communities*. Here, individualization is considered as one of the crucial driving forces. An interesting line in urban sociology attempts to explain increasing individualization of urban spaces as a sort of counter-reaction to the exteriorizing and rationalizing practices in macro-level economic and social relationships (Beck et al. 1994). Individuals are in search of new spheres of privacy and intimacy—specialized domains of "belonging to" and "safe havens" in a modernizing and place-detached world (Ascher 1995; Bourdin 2000, 2005). This reflects on the need to create new safe havens within an urban context, because it is herein that familiar ties of social integration are being eroded. Individuals aim at new appropriation of urban spaces and the "gating of communities" among their own group, as a way of ensuring that they are well protected against the outside world. As a consequence, there is a strong tendency towards the development of small city quarters or neighborhoods of internal homogeneity in different gated communities—bringing about a region-wide mosaic of differentiated living environments. The tendencies to a selective development of local environments are to be found not only in the new expansive areas in the urban periphery but also in the recomposition of existing city quarters. In many European cities, even the historic inner core of the urban region is transforming, from the evident epicenter of all specialized urban activities into a far more selective urban space. These spaces become specialized places for the production and consumption of culture, providing for an urban visitors' economy. A very selective category of inhabitants in these types of historic city centers tends to lead away from the centers of social diversity so much idealized only forty years ago by Jane Jacobs (Jacobs 1961). Of course, there is wide local differentiation within this respect, but the general tendency toward specialization of urban milieus is a general one. Alain Bourdin analyzed the next forms of recomposition of social habitat in France:

- patrimonial NIMBYs (grouped owners defending the patrimony and assuring social homogeneity within their community);
- thematic territories (focused, for instance, on leisure, sport, or age);
- ethnic territories (enclaves of ethnic groups) (Bourdin 2005).

Obviously, the creation of new spaces of intimacy is not necessarily bound by place. Processes of social dynamics in cities are complex, and there is a permanent need for further empirical testing. The fascinating thing in these lines of research is that the meaning of territorial place is no longer considered as a constant. The notion of territory is no longer fixed as local evidence, but rather as a fragile and continuously rescaling condition that must be conquered and reconquered time and again (Bourdin 2005).

What is the impact of macro- and micro-level dynamics of social and economic rescaling on the spatial configuration of cities and urban regions in Europe? There is a strong differentiation of the changing urban landscape, but we think there is wide agreement on some general tendencies:

- the growth of urbanized versus non-urbanized areas, particularly in urbanized trans-European corridors (growing population, more jobs, larger urban spaces);
- the increasing significance of regional economic competitiveness in international relationships;
- the strong differentiation—if not hierarchical relationships—of regions in global social and economic networks;
- the tendencies of decentralization and polycentrism within widening borders of metropolitan regions;
- the decreasing hierarchy between urban core and urban periphery within the metropolitan regions (specialized economic services move to urban edges and places of high external accessibility—near airports, crossing highways—leading to the overlapping of urban and rural activities in the emerging "in-between cities" and social activities in the previous periphery) (Sieverts 2003);
- the fragmentation of specialized urban spaces for economic production (specialized sites for different sorts of economic production ranging from office sites, theme parks to the creative economy in the fine historic fabric, and specialized urban spaces for consumption—different shopping centers, recreation, entertainment, sport, etc.) (Gospodini 2003; Hutton 2004);
- the sorting out of specialized economic and social spaces in urban regions that may generate new forms of spatial polarization and unbalanced ecological conditions (Graham and Marvin 2001);
- the overlap of urban and rural landscapes (emerging fragile interrelationship of landscaping cities and citifying landscapes).

The dynamic change of cities in Europe brings new challenges to the political agenda. The recapitulation of tendencies of urban development demonstrates that the typical spatial balance of economic, social, and ecological characteristics in the stage of urban compactness has been disturbed by recent tendencies of rescaling, modernization, and individualization. The question is: how can strategies of collective action deal with these complex conditions in order to create new patterns of integrated and more balanced urbanization? Considering the political agendas of cities and urban regions in Europe, almost all urban regions are wrestling with this enormous challenge. Given the dynamic context of multi-actor and multi-level governance, it appears to be quite difficult to find successful responses in the current stage of urban transformation (Newman and Herrschell 2002; Salet et al. 2003; Buck et al. 2005). This also goes for the planning strategies in the Dutch Randstad. We will discuss this particular case in more detail, certainly not as an example of "best practice" but rather to explore the complexity of current planning experiences in rescaling regional practices.

The Randstad case

With more than 6.5 million inhabitants, the Randstad is the most urbanized part of the Netherlands. The horseshoe-shaped area is only 50 miles long and 40 miles wide, but it comprises the four largest cities of the country: Amsterdam, Rotterdam, The Hague, and Utrecht. As the central area largely consists of wetlands, the Randstad is often imagined as a "greenheart metropolis." The metaphor Randstad (Rim City) suggests the existence of a coherent city. In reality, the social, economic, and administrative boundaries are far more fragmented. There is no such thing as a Randstad administration; it is just a collection of 173 municipalities, seven urban agglomerations and the urban parts of four provinces. The whole fabric is loosely coordinated by light forms of mutual consultation and cooperation between local representatives, the four provinces, and the government. The labels "Randstad" and "Greenheart" have been used for more than forty years in the national memorandums on spatial planning, but the four major city agglomerations and the additional smaller urban agglomerations have never produced a joint bottom-up spatial plan. The underlying urban patterns have always been decentralized, reflecting the polycentric history of relatively autonomous small- and medium-sized cities. Facing the rivers, the polders, and in particular, the sea, the cities have been building forward on their legacies of outward-bound commerce and trade since the golden age in the seventeenth century.

Instead of industrial production and technology, the Randstad economy specializes in international trade and advanced services. Typically, the economy is to be characterized as a transit economy. Airport Schiphol—the most important source and generator of employment in the northern Randstad conurbation—increasingly relies on transfer passengers (42 percent in 2006) (KLM et al. 2006). The relatively small domestic market makes its position a primary hub for Air France-KLM and SkyTeam partners dependent on its interconnectedness with external networks of cities. The trump card of Schiphol is its superb (internationally leading) intercontinental network with more than 100 destinations. The main threat to its future is the concentration of the SkyTeam network in Paris, the home base of leading partner Air France (KLM et al. 2006: 121). In light of this, Schiphol must produce higher quality and more efficient services, in order to maintain its status as the primary hub for SkyTeam (this has been the case in the first few years of an integrated SkyTeam). The extremely competitive position of the airport urges Schiphol not only to improve the quality and efficiency of the airside connections but also to take a strategic position in landside infrastructures and regional spatial planning. New service economies are connected with the airport. Fascinating examples of the globalizing transit economy are the trade in flowers and vegetables, and the operation of so-called "green ports." Flowers and bulbs used to be grown in the sandy parts of the Randstad. Nowadays, more than 35 percent of the roses are produced in Africa (particularly in Kenya), imported to the Netherlands, and then exported again. Cut flowers are produced and consumed at different places around the world, but amazingly, more than 60 percent of the international trade in cut flowers and 40 percent of indoor plants is traded via the auctions in the Netherlands, and is usually distributed via airport Schiphol (www.bloemenveiling.nl). Within one generation, the production of flowers has changed from simply production and export into a sophisticated trade and services economy with highly specialized forms of knowledge and employment, enabling an advanced international position.

The concentration of financial and legal headquarters at the southern ring road of Amsterdam (Amsterdam Zuidas) is facilitated with high-level international accessibility (air, rail, and highways), thus enhancing the position of the airport (Salet and Majoor 2005). Financial and legal services used to be situated in the historic center of Amsterdam, and city planners attempted to keep them there in the early 1990s by planning a mega site in the heart of the city. Instead, the major banks moved to the southern periphery, which started the development of a completely new economic center at the southern ring road. Since then, it has evolved into the most prestigious office site of the Netherlands. The financial and legal complex is therefore highly connected with international networks, as is the third economic complex of the Randstad —the highly internationalized cultural economy, the knowledge specialization, the political and legal specializations, and the visitors' economy in different urban centers of the Randstad. These forms of cultural economy typically take hold in the historic parts of cities.

Another major economic complex of the Randstad is the transportation and distribution hub of Rotterdam seaport. The port of Rotterdam is the most important source and generator of employment in the southern conurbation of the Randstad. It serves the European hinterland, not least the industrial regions of Germany. Besides chemical processing of incoming oil and gases, most incoming goods are directly transshipped and transferred as fast as possible to new destinations. The current regional, spatial, and economic planning is not prepared for unpacking the containers and processing goods but for optimizing the transfer of goods (Stevens 2000). If the goods were processed on site, added economic value and employment would likely be created. These activities could potentially lead to the sophistication of transit functions, generating other forms of economic specialization with value-added potential. The huge volumes of transport over sea enable the creation of high economic potential, not only for logistics and distribution-oriented economies but also for international trade and specialized services that guide the processes of transfer (such as finance, insurance, and legal services).

Considering the whole of the Randstad, there is no evidence of increasing completeness based on differentiated specialization. The northern and the southern conurbations still maintain different economic profiles, but there is a certain tendency toward convergence, as the same sorts of commercial services are expanding (Meijers 2005). Still, the northern part of the Randstad, particularly the Schiphol–Amsterdam zone including Amsterdam Zuidas (the concentration of financial and legal headquarters at the southern ring road of

Amsterdam), tends to concentrate the commercial service activities at the highest level of specialization—the headquarters, the most advanced international services. (Oort et al. 2006; Lambregts et al. 2006). This particular area has grown into the epicenter of the national economy.

The dependence on international connectivity is both the strength and the weakness of the Randstad transit economy. This economy is extremely vulnerable to changes in spatial-economic networks. One might wonder, for instance, whether it might be a bit artificial to let flowers from Africa be auctioned in Amsterdam in order to be sold in New York; to let passengers from Scandinavia transfer in Amsterdam in order to be transported to Southern Africa; to let computer equipment from Asia be transshipped in Rotterdam in order to get it in France. Serving as the gateway to Europe and lacking the power of a large domestic market, the Dutch economy has specialized in being actively connected in international economic and social flows of goods and information. The risks cannot be neglected; they are manifest and overt in all the examples above. On the other hand, over time, the processes of learning have resulted in an institutional capacity that enables the economy to cope with these sorts of risks and to capitalize on the transaction costs of transferring goods, people, money, and information. The highly competitive position of Schiphol airport, the flower auctions "Flora Holland" in the Randstad, and the seaport of Rotterdam urge these corporations to perform at a high level of quality, with advanced and specialized knowledge, and a minimum of costs. This may explain the active involvement of Schiphol airport in all sorts of relevant activities beyond the direct functional scope of the port authority, such as the involvement in the strategic spatial planning of neighborhoods, the landscape, and so on. This active involvement is not evident for other corporations, although we found evidence of it in the Randstad region. (Compare the relative indifference to issues regarding the regional context of airports with such large hinterlands as Heathrow in London or Kennedy Airport in New York.) The new future vision of the airport contains sophisticated ideas about the optimization of the region (KLM et al. 2006). Obviously, corporatist organizations operate in these policy networks with their own bias and interest (only the environmental lobbies manage to organize an effective countervailing power against these corporate interests). However, for the analysis of metropolitan governance, it is important first to conclude whether corporations participate in the networks of collective action at all. Then their role has to be analyzed in more detail.

The social composition of the Randstad population is highly transnational. The highest rates of migrants are to be found in the major cities, particularly in the postwar social housing quarters. Amsterdam is the most international city in the Randstad, with 48 percent of its inhabitants of foreign origin. In the wider region of Amsterdam, the percentage drops to 36 percent (Department of Research and Statistics 2008). In Rotterdam, the figures for the city and the city region are respectively 44 percent and 32 percent (CBS 2008). Most migrants rent a home from the social housing associations. Again, we can see

Table 3.1 Housing stock in the Netherlands (number of dwellings)

	1975	*1980*	*1985*	*1990*	*1995*	*2000*	*2004*
Total housing stock	4,388,002	4,849,719	5,384,081	5,892,241	6,276,045	6,650,911	6,809,581
Privately owned (%)	37.1	40.8	42.7	45.4	48.5	52.7	55.7
Rented (%)	62.9	59.2	57.3	54.6	51.5	47.3	44.3

Source: CBS 2006.

Table 3.2 Objectives and realization of housing construction in the Netherlands 1995–2005 (number of dwellings)

	Objective	*Realization (net)*	*Realization in %*
Total VINEX Area	460.4	417.64	91
Total (Netherlands)	650.4	680.39	105

Source: CBS 2006.

the differences between the high level of social housing in the cities and lower levels in the city region. Although the proportion of social housing has been decreasing since the early 1990s in favor of the private home sector, social housing associations are still the largest category in the housing markets in the cities (55 percent of housing stock in Amsterdam and Rotterdam) (see Table 3.1).

The proportion of social housing is much lower outside the major cities and city regions. For this reason, the first generations of relatively poorly educated migrants are usually dependent on housing within the boundaries of the major cities and in some of the surrounding medium-sized cities. At the higher end of the housing market, the availability of private homes within the cities is so scarce that prices for these types of properties are heavily inflated (far more in gentrifying Amsterdam than in Rotterdam), making private homes less attainable for the middle classes. Furthermore, in the major cities—particularly Amsterdam—the largest segment of households consists of one or two persons. Families with children move to the outer fringe as soon as they can afford a private home.

The combined urban and national spatial policies of urbanization promoted the compact extension of cities through the planning of new residential quarters within existing cities and in nearby urban areas. In the early 1990s, the national memorandum on spatial planning (the VINEX report) decided to allow the construction of a total of 460,000 new dwellings in these so called VINEX areas between 1995 and 2005 (Ministerie VROM 1990). This objective was successfully (91 percent) implemented, although the government could not prevent the simultaneous construction of unplanned private homes in non-designated areas (see Table 3.2).

Initially, the government pursued a spatial policy of compact urbanization for environmental reasons, but the public infrastructure investments in the VINEX areas were lagging behind. As a result, the use of private cars was much higher than expected. The VINEX areas within the city boundaries generated 49 percent car owners (on average); in the outside areas, these proportions climbed to 67 percent (Snellen et al. 2005). In a new national memorandum, the Fifth Report on National Spatial Planning, the government introduced a new concept of urbanization in lieu of compact expansion—urban networks (Ministrie VROM 2001). The government had become aware of the dynamic urban development beyond the VINEX areas of compact expansion, and launched an innovative concept of urban networks in response to increasing polycentrism. Here, a new balance between economic, social, and ecological tendencies was sought out at the regional level through the design of networks of urbanization, characterized by organized spatial relationships between nodes of urban condensation. This was an interesting new concept, albeit not yet fully supported nor elaborated in extensive programs such as the previous VINEX plan. Furthermore, in the final stage of decision making, the new (political conservative) government preferred to adapt these proposals in a new plan and opted to decentralize decisions on urbanization to local and provincial authorities (Ministrie VROM 2005).

To conclude, the rescaling of social and economic activities via globalization and modernization at the macro-level, along with individualization at the micro-level of residential areas, is resulting in a new, and as yet, not quite crystallized spatial configuration of the Randstad. Specifically, this entails more decentralized, more polycentric, and more specialized places for different sorts of economic activity, as well as the sorting of daily living milieus in a regional mosaic. The traditional hierarchy of the urban core and its periphery is changing, in more complex patterns of polycentrism. The familiar model of compact urban expansion is evolving into a sort of in-between landscape of merging urban and rural activities. Although the government initially responded with the innovative concept of urban networks, it has not been continued. In the contemporary dynamic and rather disruptive spatial context, new issues are presenting themselves to the political agenda concerning the striking of a greater balance between economic, social, and ecological tendencies. This might be enhanced by shaping new urban nodes at the level of regional networks, improving network quality by the creation of regional public infrastructure (well related with spatial development), creating new qualitative cultural urban/regional landscapes, and preventing the splitting of prosperous and back-lying spaces. These challenges of spatial planning have to be met in a new and highly dynamic context of governance. We will discuss the contextual change of governance in more detail.

The changing context of governance

Organizing collective action in regional networks is no longer an exclusive issue for local or regional planning authorities. Earlier, we observed that primarily, processes of rescaling of social and economic activities in the private sector may be considered as the main drivers of changing patterns of cities. These private sector networks are characterized by a high variegation and are often strongly interrelated with external domains (see Figure 3.2). In functional economic networks, decisions over local behavior have to be considered within the external functional domain of specialized markets, such as financial or cultural economic networks. When the major Dutch banks in Amsterdam decided to move their headquarters out of downtown Amsterdam, the decision had a dramatic impact on the position of the city center: it was no longer the historic financial center of the Netherlands. The decision was against the wishes of the city planners, who were in the midst of planning a new center for financial and commercial services in the heart of the city. Although the major banks were Dutch owned, it was not merely a local decision. The management of the corporations had to decide within the margins of the financial economic networks as to how to bring Amsterdam in direct competition with London and other international financial centers, and this is why the city of Amsterdam soon agreed to move their headquarters to the more suitable location in the southern periphery, near the international airport. This type of non-place-bound decision on place-specific choices is made in all parts of the private sector, from networks in economic markets to social organizations (including schools and housing associations, civic groups, and interest associations). As a consequence, planning strategies have to deal with stakeholders or spokesmen in this diversity of social and economic non-place networks. The second type of interrelationships consists of a variety of public sector agencies operating within the metropolitan region: the 173 municipalities, the seven agglomerations, the plethora of inter-municipal arrangements on different sorts of policy, a number of water authorities, and the four provinces within the Randstad region. This fabric of internal metropolitan actors in the public and semi-public sector is characterized by high complexity and numerous formal and informal interrelationships. Next, as the third general category of relevant interrelationships in the new context of governance, one must make mention of the interregional and trans-regional networks of public policy. In the Netherlands (like everywhere else in Europe) the nation-state took a very active role in arranging urban policies in the building up of the national welfare model from 1950 to 1980. Within that period, the national government was the most decisive actor in the public sector on critical issues such as strategic urban development and the production of social housing, the programming of public transportation, and the production of social amenities (schools, hospitals, recreation facilities, etc.). About 90 percent of local public expenditure was provided in one way or another by the central government. In most cases, the government specified the conditions for expenditure in detail. Even private organizations, such as housing associations, schools, and hospitals, were dependent on the national government. Since the early 1980s, this central–local axis has changed dramatically, due in part to the liberalization of public policies (housing, health, education), and also because of changing intergovernmental relationships. Many policy competences were decentralized to regional and local authorities, as well as to functional organizations. In addition, other governmental competences have been reorganized via international agreements such as the European Union. Thus the arena of intergovernmental relationships is extremely dynamic and very powerful with respect to regional policies. New policy games between public bodies and private sector organizations have become feasible across different levels of scale; the power of collective action for metropolitan planning strategies is arranged via completely new networks of organization. Regional organizations may forge coalitions with other regions, cooperate with national states or Europe, or organize a countervailing power jointly with Europe against the national state —which happened numerous times in Germany. All sorts of inter-scalar games have become feasible in this dynamic context of governance.

An indicative scheme may summarize the new context of multi-actor and multi-level governance in a very schematic way.

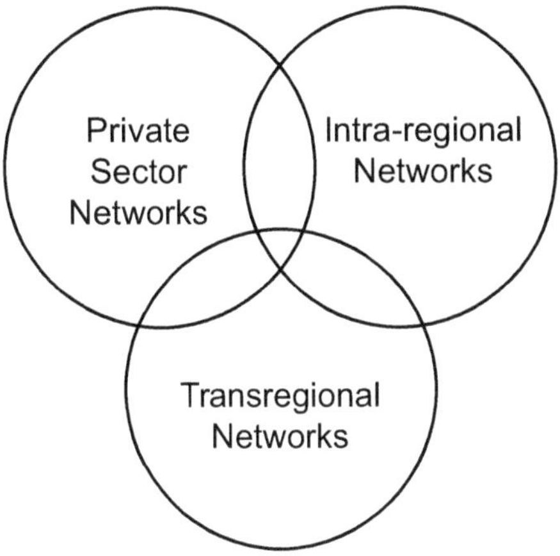

Figure 3.2 Three domains of metropolitan action

Strategic planning has become something completely different in the new dynamic context of governance. Before, public planning authorities were used to acting as nested institutions in charge of the spatial development in their territory. The familiar discussion regarding the performance of local or regional planning authorities dealt with the designation of policy competences ("central government should allow local planning authorities more resources") or regarding the level of scale of jurisdiction ("expand the area of jurisdiction of local planning authorities" or "administrative regional reform"). However, under the new conditions of governance, completely different questions are at stake in order to understand the legitimacy and efficiency of metropolitan planning strategies (Le Galès 1998). The crucial question is in the positioning of planning arrangements, in the different domains of metropolitan action. Superior strategies are those that manage to optimize their position in networks within and between the three domains of action in Table 3.2. Spatial planning no longer depends on the gravity of nested territorial planning authorities but on the intelligence and flexibility of interconnected relationships (Salet et al. 2003).

The challenge to address new regional questions in this highly variegated fabric of governance is a complex one. It requires strategies of intermediation that are difficult to fulfill due to differing interests and ways of dealing with uncertainty. The rhythms of public and private sector agencies are very different. In the Randstad, the development of various private sector networks, particularly internationally relevant economic services and cultural networks (e.g. architecture, lifestyles, and musical expressions), took the lead in changing local patterns. The governmental agencies, including local and regional planning authorities, were not proactive in the changing patterns of urbanization. Previously, they made the spatial plans for city and region, but they are no longer the main drivers of change. The city evolved dynamically; at best, the planning bodies were not counterproductive (as happened in the example of the financial heart of the city referred to above), and sooner or later managed to accommodate the vibrant social and economic trends. In a sense, the rhythm of private sector networks is more dynamic than that of public sector networks. They may cross the territorial scales of city and region more easily and swiftly than nested bodies of government and planning. They do so, however, within the very narrow lines of their functional scope. The activities of financial institutions, airports, and cultural activities are completely interrelated with external global sources, but only with respect to their specific activity. Schiphol airport is one of the few corporations with a more integrated view of regional development, whereas major banks focus mainly on hot spots of development.

Public sector agencies in the Randstad, on the other hand, are rather introvert, their scope is only gradually becoming outward looking. There is a lot of unrealized potential in the arrangement of cooperating or countervailing alliances in external public policy interrelationships. Many public agencies within the region still play the local cards of the previous epoch while the new context of the game should be urging them to play the external cards more actively. On the other hand, public agencies are typically interested in the integration of different sector policies. Public planning agencies in the Randstad are proactive in integrating spatial policies with sector policies on infrastructure, water, housing, development of economic sites, and the creation of landscapes, among other issues. In this respect, private sector networks—even civic groups—are usually far less interested and less equipped with resources. The differences in rhythm between public and private sectors require a lot of intermediation and inter-sector learning strategies.

Empirical findings of research into the most strategic projects in the Randstad may illustrate these differences. The Amsterdam Zuidas office district has grown in the last ten years into the most advanced international center of commercial and legal services in the Netherlands. When the major banks in the country decided ten years ago to settle in this strategic position between the airport and the city, the municipal planners had to agree with their downtown exodus, and they followed suit a few years later with new plans for the development of an integrated urban center, in Zuidas (aiming at a "second center" of the city). The market actors operate in international networks, expanding and reducing all sorts of local activities according to the logic of their specific markets (e.g. increasing or decreasing the number of employees at the site, expansion or reduction of office space). During the recent financial crisis, the major ABN/AMRO bank was sold to an international consortium (and later taken back by the Dutch state). Their interest in development is primarily functional in scope—prestigious offices, high standards for the site, accessibility, parking garage. They make no further claims with respect to the integration of the site in the urban and regional fabric, with the environmental quality of the site in the region, public spaces, social housing, etc. The spatial planners act at a certain distance from the dynamic international market players, not being informed of all the "ins and outs," but gaining profit from the expansive epoch. The planners jumped in at the local level, bringing a strategic plan constructed with ambitions to integrate local spatial qualities (different kinds of social, economic, and cultural activities in Zuidas). Over the last eight years, they have gradually managed to enlarge the scope of public action. The central government got involved by financing a portion of the infrastructure. The region got involved by recognizing the area as a site to compete with other potential sites in the region. Even under the conditions of the present financial crisis, the level of public investment has been stabilized. So far, the Dutch government has stabilized the level of public investment in order to support the private sector. However, the urban planners have not yet managed to get the Zuidas project to the top of international policymaking agendas. Although considered one of the most competitive projects in Europe, planners are still attached to the familiar bottom-up networks and their contacts with the national government, rather than new interregional and international lobbies and policies. Why not actively market in international milieus, like the growing Chinese economy? Why not bring the project to the competitive Lisbon agenda of the European Union? Why not make it the most environmentally friendly project in Europe, a model project of the European Union, searching at this level for the enhancement of balancing issues, instead of purely economic competitiveness? Why not establish Zuidas as an international center of culture, as demonstrated in the Barcelona Forum project (Salet and Gualini 2006; Majoor 2008)? In practice, this strategic project was developed during its first ten years solely as a commercial development. In this respect it was—at least before the financial crisis—more successful than in other respects. Although the market and public sector planning bodies are closely interrelated, other potential participants may find it difficult to get involved (Majoor 2008). This way of planning is still dominant. Arranging interconnectivity between the different domains of metropolitan action might bring more intelligence and flexibility in the actual context of governance. We qualify the present style of planning mega projects in the Randstad as learning and only gradually improving experiences in the current process of urban transformation.

The role of power in metropolitan governance

In the context of Randstad governance, the institutionalization of power is one of the constraining forces in the current process of urban change.

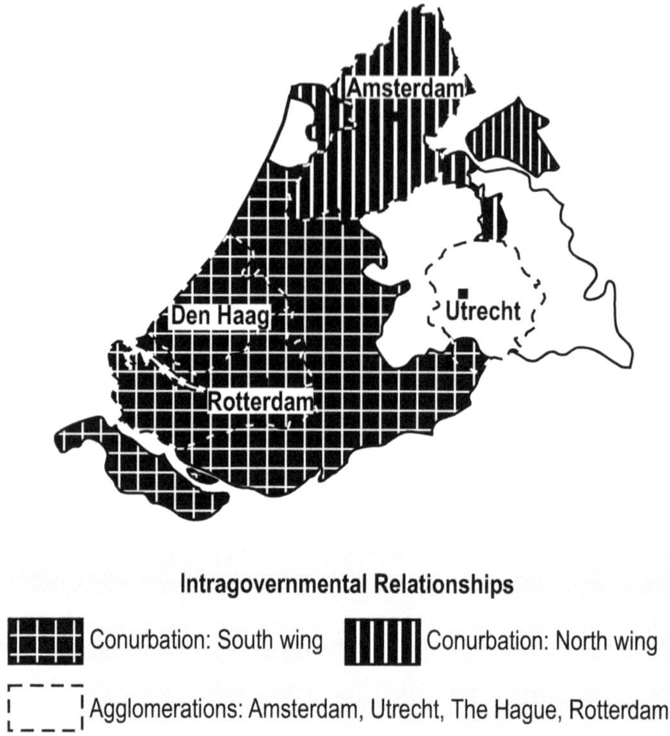

Figure 3.3 Agglomeration and conurbation in the metropolis of Randstad
Source: Chapter author.

The transformational power of institutions is not yet fully utilized in the present transition of cities in higher-level configurations of urbanity. In the establishment of planning strategies, by far the most energy consumed is in the internal relationships of the intra-regional networks of public sector agencies. The 173 municipalities and other public agencies put a lot of effort into creating mutual strategies of cooperation at various levels of scale. This circle of "intra-regional networks" is by far the most frequent in the creation of metropolitan planning strategies. Considering spatial development, one might make an indicative distinction between three levels of regional configuration in the Randstad: (1) the level of urban agglomeration; (2) the level of urban conurbation; and (3) the metropolitan level (see Figure 3.3).

The first level—"urban agglomeration"—consists of a major city and its surrounding suburbs. Region Amsterdam consists of the city of Amsterdam and sixteen surrounding municipalities. They cooperate on a voluntary basis. Rotterdam, The Hague, and Utrecht also have these kinds of urban agglomerations. There are additional agglomerations around medium-sized cities in the Randstad. The second level—"urban conurbation"—brings together several urban agglomerations. The north wing of the Randstad is a conurbation overarching the agglomerations of Amsterdam and Utrecht. Typically, a lot of overlapping activities occur in this urban conurbation. However, the creation of successful cooperation between the two agglomerations appeared to be unfeasible. In the southern wing of the Randstad, there is an urban conurbation, including the agglomerations of Rotterdam and The Hague, along with some smaller urban agglomerations. The policy interrelationships in these conurbations are also based on voluntary agreement. The third spatial configuration is the full "metropolitan level" of the Randstad. Since 1958, the Randstad has been mentioned in all national reports regarding spatial planning. However, it also relies on voluntary association. Since the mid-nineteenth century the formal administration has been organized at the different levels of municipalities and provinces. This is why all three forms of regionalization strongly depend on mutual cooperation between municipalities and provinces.

Housing, labor, and the corresponding mobility are most intensively played out at the level of urban agglomeration in daily urban systems. At this level, we may find the highest proportion of daily home–work trips; housing and labor markets also operate at this level of scale. The significance of the "urban conurbation" is strongly increasing for the same type of activities, due to the increasing overlap of urban agglomerations. For instance, the majority of the inhabitants of the Rotterdam VINEX neighborhoods are employed in The Hague, and many residents of the Utrecht VINEX location Leidscherijn are employed in the Amsterdam agglomeration. The "metropolitan" configuration is not yet an urban configuration, although there is cooperation at this level with respect to external marketing. At this level, a new, fascinating agenda is emerging—not on specific urban issues, but on concerns regarding water, landscaping, and living beyond the urban agglomerations.

It is fascinating to consider the power relationships in the production of planning strategies in this complicated urban and regional fabric. Private sector networks, from economic networks to civic groups, are generally not involved in the production of regional planning strategies. Planning is the core business for cooperating and conflicting public agencies. The most powerful stakeholders are the four major cities, along with the government. Provinces are not very powerful in these urbanized parts of the country. Major cities do not like to be interfered with by provinces, and they keep their own contacts with national ministries. The power of the major cities manifests itself at the level of urban agglomeration and via informal consultancies; the major cities also have a strong voice in national policies. At the level of urban agglomeration, the major cities are dominant. They are better equipped and carry more planning expertise than smaller surrounding municipalities, and are more represented in external relationships, considering the agglomeration as their own backyard. This power asymmetry has generated distrust at certain times, but the developmental agenda of the last eight years has shaped very cooperative minds. The major cities have successfully managed to get the agglomeration's interests recognized in national urban policies. Although public infrastructure investment is lagging behind (there is still a sharp boundary between public infrastructure within the city boundaries and public infrastructure within larger urban agglomerations), new investment is gradually moving to the level of urban agglomeration.

The higher level of "urban conurbation" is more complex, when considered from the perspectives of the major cities. The agglomerations overlap so much that it is necessary to think strategically about urban networks at this level. However, the major cities at this level have no strong interest in policy-making. They find it difficult to share their privileged position with other urban agglomerations. Also, the interests of the electorate keep urban politicians in their "own" urban scenery. There exists some rivalry, as urban conurbations are expanding in the same housing and office markets, the same arenas they wish to attract to their own cities. Rotterdam, for instance, opposed a middle-class extension in Gouda (25 miles from the city) because of similar plans in their own city. Why should Rotterdam favor the settlement of interesting theme parks in the "in-between cities" of the urban conurbation rather than its "own" city or agglomerations? There are cooperative policy networks at the level of conurbation, often co-chaired by the major cities, but their aspirations lie just a bit more at the level of agglomeration. After a few years of consultation and cooperation between the agglomerations of Amsterdam and Utrecht, they decided to focus on their own agglomeration once again. There is no policy cooperation at the level of urban conurbation in Randstad North. In Southern Randstad, there is still consultation between the relevant agglomerations via the "South Wing Randstad Platform." Unfortunately, the platform is not known for effective decision making. The province is the most involved actor in this strategic platform, regularly bringing new issues to the agenda. The major cities, on the other hand, pay lip service to the platform and keep a low profile. Hence there is a mismatch between the dynamic spatial developments and the administrative forms of policy cooperation. The agglomerations of The Hague and Rotterdam overlap, and the inhabitants cross the administrative boundaries frequently, but the inter-municipal cooperation remains centered around the two urban centers.

The same concept applies to the higher levels of metropolitan scale. The Randstad was invented and is recognized by the national planners, but is not recognized by the cities themselves. For matters of city marketing abroad and within the national government, the involvement of lobbies is useful and effective. It keeps urban matters on the political agenda of the national government. The real social and economic coherence, however, is at a much lower level. With respect to politics, the major cities pay lip service to the metropolitan region, but the actual commitment is rather low. When Amsterdam learned last year that its own agglomeration was the real point of gravity in the national economy (including the entire Randstad), it even started to label the strategic policy of its own agglomeration as the metropolitan region of Amsterdam! The bottom-up urban profile of the Randstad is rather low, but interestingly, a new political agenda is emerging at the Randstad level. It is not a typical urban agenda, but water has become the most critical issue in the metropolitan planning strategy. Issues concerning water (change of climate, higher sea level, the salinization of freshwater, the landscape, etc.) have put new adaptive water policies on the agenda. Here, new issues of integrating physical water issues (e.g. enlargement of rivers, inundation of polders, design of new water basins, guarantees for freshwater) with landscaping and innovative milieus for living on the water, are emerging on the political agenda. So, at the metropolitan level, a different sort of agenda is unfolding, which should not be considered as an urban agenda but as a mixed agenda of ecological, economic, and social trends, in a wide urban/rural setting. What we gradually begin to learn is that the metropolitan region should not be considered as an expansion of cities but as a new combination of rural and urban, with its own need of enhancement in powerful strategies.

Conclusions

What can we learn from the Randstad experiences with planning strategies in a dynamic context of governance? The rescaling of social and economic parameters generates an ongoing process of decentralization, increasing polycentrism, and specialization of urban spaces. The dynamic private sector developed its own action spaces in the expansive urban system through the 1990s. This was the case for both economic development and residential areas beyond the VINEX plans. The current planning strategies are not yet completely responsive to the change of social and economic parameters. The dynamics of urban transformation are rooted in the private sector, but the planning strategies are created by the public sector. There are some mismatches between the two domains. More intermediation between these different experiences will undoubtedly improve the intelligence of planning strategies. The external (trans-regional) orientation of public sector planning agencies is still in its embryonic stage. Considering the three domains of metropolitan action, the current planning strategies are mainly produced in the internal domain of local and regional public sector agencies.

Three types of new regional configuration have been described: urban agglomeration, urban conurbation, and metropolitan networks. The planning strategies for responding to these three types of regionalization in the present context of governance are dominated by the four major cities and the government. As a result, most of the attention goes to strategies at the level of urban agglomeration. The financial crisis has not yet changed this dominant pattern. The cooperation between municipalities at the level of urban agglomeration has become relatively successful; policies are increasingly integrated within agglomerations and effectively lobbied by the central government. However, in the actual spatial tendencies, the level of conurbation is becoming more significant, but the power and prevalence of the major cities frustrate the production of effective strategies at this level of scale. Finally, at the level of the full metropolis, we may find that a new policy agenda is emerging.

This agenda should differ strongly from those of the commercial interests and governments of the major cities. However, this new agenda is still in search of power and effectiveness. The metropolitan region is more than an extension of the urban level. Planning strategies for the metropolitan level should not be made dependent on the commitment of the conservative cities. They should be distinguished from urban-level strategies. The neglected provinces might take the initiative for new metropolitan planning strategies on emergent, non-urban planning issues regarding the integration of water, landscaping, and residential policies, and create their own networks of social endorsement.

References

Amin, A. and Thrift, N. (2002) *Cities: Reimagining the Urban*, Cambridge: Polity Press.

Ascher, F. (1995) *Métapolis ou L'avenir des Villes (Métapolis or the Future of the City)*, Paris: Odile Jacob (in French).

Beck, U., Giddens, A., and Lash, S. (1994) *Reflexive Modernisation: Politics, Tradition and Aesthetics in Modern Social Order*, Cambridge: Polity Press.

Bourdin, A. (2000) *La Question Locale (The Local Question)*, Paris: Presse Universitaire de France (in French).

—— (2005) *La Métropole des Individus (The Metropolis of Individuals)*, Paris: l'Aube (in French).

Brenner, N. (2004) *New State Spaces: Urban Governance and the Rescaling of Statehood*, Oxford: Oxford University Press.

Buck, N., Gordon, I., Harding, A., and Turok, I. (2005) *Changing Cities*, New York: Palgrave.

CBS (Centraal Bureau voor de Statistiek/Netherlands Central Office of Statistics) (2006) *Statline: Woningbouw, VINEX en Nieuwbouwwoningen (Housing Construction Fourth Extra Memorandum on Spatial Planning (VINEX) and New Housing)*, Den Haag/Heerlen: CBS (in Dutch).

—— (2008) *Statistisch Jaarboek 2008*, Voorburg/Heerlen: CBS.

Department of Research and Statistics (2008) *Metropoolregio Amsterdam in cijfers 2008*, Amsterdam: Municipality of Amsterdam.

ESPON (2006) *ESPON Atlas: Mapping the Structure of the European Territory*, Bonn: Federal Office for Building and Regional Planning.

Fishman, R. (1987) *Bourgeois Utopias: The Rise and Fall of Suburbia*, New York: Basic Books.

Gospodini, A. (2003) "Portraying, Classifying and Understanding the Emerging Landscapes in the Post-industrial City," *Cities*, 23(5): 311–30.

Graham, S. and Marvin, S. (2001) *Splintering Urbanism*, London: Routledge.

Hall, P. and Pain, K. (eds.) (2006) *The Polycentric Metropolis: Learning from Mega-city Regions in Europe*, London: Earthscan Publications.

Hall, P., Thomas, R., Gracey, H., and Drewett, R. (1973) *The Containment of Urban England*, London: Allen & Unwin.

Hutton, T. (2004) "The New Economy of the Inner City," *Cities*, 21(2): 89–108.

Jacobs, J. (1961) *The Death and Life of Great American Cities*, New York: Vintage Books.

KLM, Air Traffic Control, and Schiphol Group (2006) "Connecting Cities," in W. Salet (ed.) (2006) *Synergy in Urban Networks?*, Den Haag: SDU Uitgevers, pp. 111–35.

Lambregts, B., Kloosterman, R., Van der Werff, M., Röling, R., and Kapoen, L. (2006) "Randstad Holland: Multiple Faces of a Policentric Role Model," in P. Hall and K. Pain (eds.) (2006) *The Polycentric Metropolis: Learning from Mega-city Regions in Europe*, London: Earthscan, pp. 137–45.

Le Galès, P. (1998) *Regulations and Governance in European Cities*, Oxford: Blackwell.

Lefebvre, H. (1991) *The Production of Space*, Oxford: Blackwell.

—— (1996) *Writings on Cities* (eds. E. Kofman and E. Lebas), Oxford: Blackwell.

Majoor, S. (2008) *Disconnected Innovations: New Urbanity in Large-scale Development Projects in Zuidas Amsterdam, Ørestad Copenhagen and Forum Barcelona*, Delft: Eburon.

Marcuse, P. and Van Kempen, R. (eds.) (2000) *Globalizing Cities: A New Spatial Order*, Oxford: Blackwell.

Meijers, E. (2005) "Polycentric Urban Regions and the Quest for Synergy: Is a Network of Cities More than the Sum of the Parts?" *Urban Studies*, 42(4): 765–81.

Ministerie VROM (Ministry of Housing, Spatial Planning and the Environment) (1990), *Vierde Nota Extra inzake de Ruimtelijke Ordening (VINEX), (Fourth Extra Memorandum on National Spatial Planning (VINEX), Part 1*). Online at www.vrom.nl (accessed December 15, 2009) (in Dutch).

—— (2001) *Vijfde Nota inzake de Ruimtelijke Ordening* (*Fifth Memorandum on National Spatial Planning*). Online at www.vrom.nl (accessed December 15, 2009) (in Dutch).

—— (2005) *Nota Ruimte (National Spatial Strategy)*. Online at www.vrom.nl (accessed December 15, 2009) (in Dutch).

Newman, P. and Herrschell, T. (2002) *Governance of Europe's City Regions*, London: Routledge.

Oort, F., Van Brussel, J., Raspe, O., Burger, J., Van Dinteren, B., and Van der Knaap, B. (2006) *Economische Netwerken in de Regio (Economic Networks in the Region)*, Rotterdam/Den Haag: NAI Uitgevers/Ruimtelijk Planbureau (in Dutch).

Salet, W. and Majoor, S. (2005) *Amsterdam Zuidas: European Space*, Rotterdam: 010 Uitgevers.

Salet, W. and Gualini, E. (2006) *Framing Strategic Urban Projects*, London: Routledge.

Salet, W., Thornley, A., and Kreukels, A. (2003) *Metropolitan Governance and Spatial Planning: A Comparative Study of European City-regions*, London: Routledge/Spon.

Sassen, S. (2001) *The Global City*, 2nd edition, Princeton, NJ: Princeton University Press.

Sieverts, T. (2003) *Cities without Cities: Between Place and World, Space and Time, Town and Country*, London: Routledge.

Snellen, D., Hilbers, H., Hendriks, A., Van Amsterdam, H., and Peeters, P. (2005) *Nieuwbouw in Bewegffffg, een analyse van het ruimtelijk mobiliteitsbeleid van VINEX* (*New Housing Construction on the Move: Analysis of the Spatial Mobility Policy in the Fourth Memorandum Extra on National Spatial Planning*), Rotterdam: NAI Publishers (in Dutch).

Stevens, H. (2000) *The Institutional Position of International Seaports*, Dordrecht: Elseviers.

Swyngedouw, E. and Baeten, G. (2001) “Scaling the City: The Political Economy of ‘Glocal’ Development – Brussels Conundrum,” *European Planning Studies*, 9(7): 827–49.

Taylor, P. (2003) *World City Network: A Global Urban Analysis*, London: Routledge.

Webber, M. (1964) “The Urban Place and the Nonplace Urban Realm,” in M. Webber, W. Dyckman, D. Foley, A. Guttenberg, W. Wheaton, and C. Bauer Wurster (eds.) *Explorations into Urban Structure*, Philadelphia, PA: University of Pennsylvania Press, pp. 79–153.

4 Strategic planning and regional governance in Europe

Recent trends and policy responses

Louis Albrechts

Introduction: setting the context

Europe is facing major developments, challenges, and opportunities that are affecting its cities and regions, either directly or indirectly. They include: growing complexity (rise of new technologies, changes in production processes, the crisis of representative democracy, diversity, globalization of culture and the economy, rising costs of energy); the financial crisis and the subsequent economic crisis; persistently uneven development; the problems of fragmentation; the aging population; and the increasing interest (at all scales, from local to global) in environmental issues (e.g. global warming) (Albrechts 2001, 2004, 2006). Moreover, this agenda is expanded by: the need for governments to adopt a more entrepreneurial style of planning in order to enhance regional, city region, and urban competitiveness; the growing awareness that some planning concepts (i.e. learning regions, knowledge communities, industrial districts, compact cities, livable cities, creative cities, multicultural cities, fair cities) cannot be achieved solely through physical hard planning; and the fact that (in addition to traditional land use regulation, urban maintenance, production, and management of services) governments are being called upon to respond to new demands that imply the abandonment of bureaucratic approaches and the involvement of skills and resources that are external to the traditional administrative apparatus.

This chapter considers four possible types of response to these developments and challenges: (1) reactive (the rear-view mirror); (2) inactive (going with the flow); (3) preactive (preparing for the future); and (4) proactive (designing the future and making it happen) (see Ackoff 1981). There is ample evidence that the problems and challenges that regions, city regions, and cities are confronted with cannot be tackled and managed adequately with either the neo-liberal perspective or the intellectual, technical-legal apparatus and mindset of traditional land use planning.

My thesis is that only the proactive reaction is truly appropriate, as it calls for the transformative practices that are necessary to cope with the continuing and unabated pace of change driven by the (structural) developments and challenges. Transformative practices focus on the structural problems in society; they construct images/visions of a preferred outcome and how to achieve it (see Friedmann 1987). So, in many places in Europe, a shift is needed from a more regulative, bureaucratic approach towards a more strategic, implementation-led, and development-led approach.

As there is no "one best or one single European way" to do strategic planning, the purpose of this chapter is to add a new dimension in terms of values, approach, and process. It therefore reexamines strategic (spatial) planning by highlighting the differences with traditional land use planning, by using views from the planning literature, from the (mainly) European practices of strategic thinking at different scale levels, and from shifts in the overall planning approach. This is done by combining theory with practical experience.

Land use planning

Aims

Land use planning, in an integrated and qualitative way, is basically concerned with the location, intensity, form, amount, and harmonization of land development required for various space-using functions—such as housing, industry, recreation, transport, education, nature, agriculture, and cultural activities (Chapin 1965; Cullingworth 1972). In this way, a land use plan embodies a proposal to show how land should be used—in accordance with a considered policy—as expansion and restructuring proceeds into the future.

A classification of EU land use planning systems

The emergence of land use planning systems across Europe has some common roots. In many European Union (EU) member states, their first planning laws were produced in the early twentieth century as a response to increasing development pressure and the consequent problems that arose from dense and disorganized development. Cultural, institutional, and legal differences, in addition to specificity of purposes, produced a wide variety of planning systems and traditions. The EU Compendium of Planning Systems and Policies (CEC 1997) draws a line between strategic planning at the regional/national level and land use planning at the level of the municipality and the functional urban region. However, practice (Healey et al. 1997; Pascual and Esteve 1997; Salet and Faludi 2000; Albrechts et al. 2001; Pugliese and Spaziante 2003; Albrechts et al. 2003; Martinelli 2005; Healey 2007; Balducci 2008) illustrates that much strategic planning is happening at the level of the city and the urban agglomeration.

Land use planning at the EU level clearly focuses on the municipality or functional urban region (mainly in Greece, France, Italy, and Sweden), with framework instruments, and on specific areas within the municipality, with regulatory instruments. The framework plans cover at least the whole of the area of the local authority and details the general land use and infrastructure patterns across the area through zoning or land allocation maps. The regulatory plan covers either part of or the entire local authority's area and indicates detailed site-specific zonings for building, land use, and infrastructure. All the EU member states apart from the United Kingdom (UK) and the Republic of Ireland use detailed planning instruments (regulatory zoning, building control, and implementation), which play a determining role in guiding the location, form, and size of development and physical infrastructure. The framework and regulatory plans are mostly legally binding documents that, once approved, have implicit legal certainty and rigidity. With the major exception of the UK, they are generally of no fixed duration and can be replaced only by new plans. This situation often relates to the extent to which a system is binding or discretionary (CEC 1997). In a binding system, it is expected that the relationship between policy and control will be determined through a binding and detailed land use plan, whose effects are compared with the original intentions in the same detailed way that a house would be compared to its original blueprint. This is the "conformance view" (Barrett and Fudge 1981). In a discretionary system, each decision is subject to administrative and political discretion, with the plan providing general guidance. This relates to the "performance view" (Mastop and Faludi 1997). The UK is the primary example of a discretionary system. Whereas in the binding system the focus is on legal certainty, there is a notable absence of certainty in the discretionary system.

Local governments are generally responsible for the production of these plans, and for most EU countries (with the exception of Greece); the responsibility for approval is also delegated to this level. Sometimes, the local plans have to be formally approved by another tier (regional or central) of government.

Consultation with other tiers of government, administrative powers, and official agencies is an essential part of plan-making. There is also a general commitment to consultation with the public; however, the method and depth of public involvement varies considerably. The Scandinavian countries have a long-standing tradition, but in other countries, it remains very formal and restricted (e.g. in Greece, Italy, and Portugal).

The approach to planning via a single policy field (i.e. spatial planning) met fierce opposition from other policy fields, which were generally more powerful. Although land use plans had formal status and served as official guidelines for implementation, when it came down to the actual implementation, other policy fields—which, due to their budgetary and technical resources, were needed for the implementation—were easily able to sabotage the spatial plans if they wanted (Scharpf and Schnabel 1978; Kreukels 2000).

A major emphasis on the legally binding nature of most EU land use plans provides official certainty, but makes the plans far more rigid and inflexible, and less responsive to changing circumstances. The mainly comprehensive nature of land use plans is at odds with the increasingly limited resources and ever-changing challenges. In addition, most land use plans have a predominant focus on "physical" aspects, providing "physical" solutions to social/economic problems. In this way, they often abstract from the real-life historically determinate parameters of human activity and gratuitously assume the existence of transcendent operational norms.

In all EU systems, there is evidence (CEC 1997) to suggest that the systems should become more open and less prescriptive in determining precise land uses in favor of a more flexible system to respond more quickly and adequately to changing social and economic circumstances and to the agenda of government reorganization.

Revival of strategic spatial planning

In a number of European countries, spatial planning evolved in the 1960s and 1970s towards a system of comprehensive planning—the integration of nearly every aspect of planning (Perloff 1980)—at different administrative levels. In the 1980s, when the neoliberal paradigm replaced the Keynesian–Fordist paradigm and when public intervention retrenched in all domains (Martinelli 2005b), we witnessed a retreat from strategic planning fueled not only by the neoliberal disdain for planning but also by postmodernist skepticism, both of which tend to view progress as something that, if it happens, cannot be planned (Healey 1997b). Within the architectural/urbanism discipline, a new approach emerged to land use regulation and urban projects (Secchi 1986; Motte 1994), especially for the revival of rundown parts of cities and regions. A new generation of strategic (mainly urban) projects, such as the French "Projet urbain," has been trying to develop a more inclusive approach, informed by insights in policy analysis and strategic planning. From these practices, a whole body of knowledge is developing, which could be described as "theorizing practice" (Masboungi and De Gravelaine 2002). However, a more theoretical framework within which these concrete practices could be framed and evaluated has not yet been developed, and the gaps remain open. Planning and urbanism seem highly complementary in their approach, as well as in their strengths and weaknesses. To construct an integrated approach there is a need for cross-fertilization between the more model-based and top-down planning views and the more casuistic, bottom-up experiences. Other discourses to be integrated concern the social, cultural, political, ecological, participative, and economic aspects. In conclusion, in both the public and private sectors, there is a need to develop a more action-oriented approach.

In response, by the end of the millennium more strategic approaches, frameworks, and perspectives for cities, city regions, and regions had once again become fashionable in Europe (Healey et al. 1997; CEC 1997; Pascual and Esteve 1997; Albrechts 1999, 2004, 2006; Salet and Faludi 2000; Albrechts et al. 2003; Pugliese and Spaziante 2003; Martinelli 2005a; Healey 2007).

As in "traditional" planning, there are different types of strategic spatial planning in Europe. This chapter focuses on what I call "new" strategic spatial planning. New strategic spatial planning provides a critical interpretation of the structural challenges and problems and thinks creatively about possible answers and how to get there. This chapter deals with new strategic planning by elaborating three interrelated components: a what, a how, and a why (see Healey 1997a, 1997b; Albrechts 2001, 2004, 2006; Albrechts et al. 2001, Albrechts et al. 2003; Motte 2006; Healey 2007; Balducci 2008).

"New" strategic spatial planning

What?

"New" strategic spatial planning is a transformative and integrative, public-sector-led, and socio-spatial process through which visions/frames of reference, justification for coherent actions, and means for implementation are produced that shape and frame what a place is and what it might become (Albrechts 2001, 2004, 2006). The term "spatial" brings the "where of things" into focus, whether static or dynamic: the creation and management of special "places" and sites; the interrelations between different activities and networks in an area; and significant intersections and nodes that are physically co-located (Healey 2004b: 46). Cities, city regions, and regions possess a distinctive spatiality as agglomerations of heterogeneity locked into a multitude of relational networks of varying geographical reach (Amin 2004: 43). Strategic spatial planning processes with an appreciation of "relational complexity" demand a capacity to "hear," "see," "feel," and "read" the multiple dynamics of a place in a way that can identify those key issues that require collective attention through a focus on place qualities (see Healey 2005, 2006). As a consequence, strategic spatial planning evolves continuously in formulation (see Healey 2007). The focus on the spatial relations of places allows for a more effective way of integrating different agendas (economic, environmental, cultural, social, and policy agendas) as these agendas affect places. As these plans have a variable reach, they also carry a potential for "rescaling" down from the national or state level and up from the municipal and neighborhood level. The search for new scales (e.g. mega-city regions) of policy articulation and concepts is also linked to attempts to widen the range of actors involved in policy processes, and to new alliances, actor partnerships, and consultative processes (Albrechts et al. 2003; see also Xu and Yeh, Hall, and Salet in this volume).

How?

"New" strategic spatial planning focuses on a limited number of key issues. It takes a "collective" critical view of the environment in terms of determining strengths and weaknesses in the context of opportunities and threats. Strategic spatial planning focuses on place-specific qualities and assets (the social, cultural, and intellectual qualities—both physical and social—of the urban/regional tissue) within a global context. It is impossible to understand material places and social nodes such as "the city," "the city region," "the region," and "the mega-city region" positioned in a one-dimensional hierarchy of scales (Healey 2007: 267). Strategic spatial planning studies the external trends, forces, and resources available. It identifies and gathers major actors (public and private) in a co-production process; it allows for a broad (multilevel governance) and diverse (public, economic, civil society) involvement during the planning, decision-making, and implementation processes. It creates realistic long-term visions/perspectives (a geography of the unknown) and strategies at different levels, taking into account the power structures (political, economic, gender, cultural), uncertainties, and competing values. Strategic spatial planning designs plan-making structures and develops content, images, and decision frameworks for influencing and managing spatial change. It provides a frame of reference that gives direction and justifies specific action. It is about building new ideas and processes that can carry them forward, thus generating ways of understanding, ways of building agreements, and methods of organizing and mobilizing for the purpose of exerting influence in different arenas. Finally, strategic spatial planning, both in the short and in the long term, focuses on framing decisions, actions, projects, results, and implementation, incorporating monitoring, evaluation, feedback, adjustment, and revision.

Why?

The "why" question deals with values and meanings, with "what ought to be." Without the normative, we risk adopting a pernicious relativism where anything goes (Ogilvy 2002). In a conscious, purposive, contextual, creative, and continuous process, new strategic planning aims to enable a transformative shift—where necessary—to develop openness to new ideas, and to understand and accept the need and opportunity for change. Transformative practices oppose the blind operation of market forces and involve constructing "desired" answers to the structural problems of our society. Normativity indicates the relations with place-specific values, desires, wishes, or needs for the future that transcend mere feasibility and are the result of judgments and choices formed, in the first place, with reference to the idea of "desirability," to the idea of "betterment" (Ozbekhan 1969), and to the practice of the good society (Friedmann 1982). To influence particular future states is an act of choice involving valuation, judgment, and the decision-making that relates to human-determined ends and to the selection of the most appropriate means for coping with such ends. This is contrary to the future as an extension of the present. The "future" must symbolize some qualities and virtues that the present lacks (diversity, sustainability, equity, spatial quality, inclusiveness, and accountability).

Four-track approach

The "new" strategic planning approach is utilized in a four-track approach. The four tracks, as shown in Figure 4.1 (Albrechts et al. 1999; see also Van

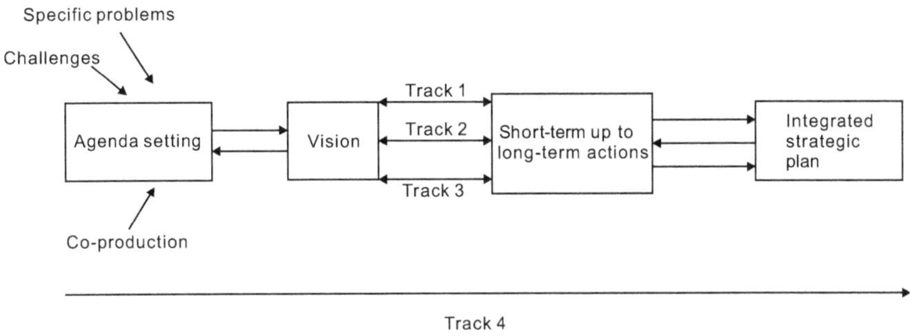

Figure 4.1 Possible macro-structure for the overall strategic planning process
Source: Chapter author.

den Broeck 1987 for the three-track approach), display the working tracks or planes (see Hillier 2007): one for vision; another for short-term and long-term actions; a third for involvement of the key actors; and, finally, a fourth track for a more permanent process (primarily at the local level) involving the broader public in major decisions. The proposed tracks may not be viewed in a purely linear way. The context forms the setting of the planning process but also takes form and undergoes changes in the process (Dyrberg 1997). The four-track approach emphasizes transformation and action. It opens up multiple potentialities immanent in space (Hillier 2005). Not every potentiality can be actualized. The selection and construction of visions and decisions on specific actions are done within a context of structural constraints.

The four-track approach is based on interrelating four types of rationality (as shown in Figure 4.2): (1) value rationality (the design of alternative futures); (2) communicative rationality (involving a growing number of actors, both public and private, in the process); (3) instrumental rationality (looking for the best solutions to achieve the desired future); and (4) strategic rationality (a clear and explicit strategy for dealing with power relationships) (Albrechts 2003a).

The end product may consist of a critical analysis of the main processes and structural constraints shaping our places, which amount to: realistic, dynamic, integrated, and indicative long-term visions (frames); plans for short-term and long-term actions; a budget; and flexible strategies for implementation. It constitutes a commitment or possible (dis)agreement between the key actors. For the implementation, credible commitments to action engagement (commitment package) and a clear and explicit link to the budget are needed where citizens, the private sector, different levels of governance, and planners enter moral, administrative, and financial agreements to realize these actions (collective spatial agreement).

Strategic spatial planning is not a single concept, procedure, or tool. Rather, it is a set of concepts, procedures, and tools that must be tailored carefully to the situation at hand if desirable outcomes are to be achieved (Zeng 2007).

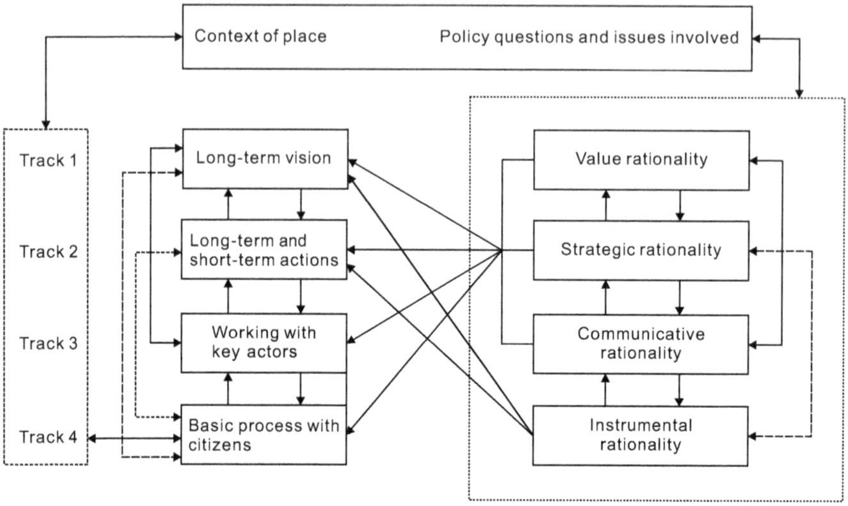

Figure 4.2 Tentative integration of different concepts of rationality
Source: Albrechts 2003a.

Strategic spatial planning is as much about process, institutional design, and mobilization as it is about the development of substantive theories. Content relates to the strategic issues selected in the process. The capacity of a strategic spatial planning system to deliver the desired outcomes is dependent not only on the system itself but also on the conditions underlying it. These conditions—including structural constraints as well as political, cultural, and professional attitudes towards spatial planning (in terms of planning content and process) and the political will on the part of the institutions involved in setting the process in motion and keeping it going—affect the ability of planning systems to (partially) implement the chosen strategies.

Strategic planning is selective and oriented to issues that really matter. It is not only a contingent response to wider forces but also an active force in enabling change. This strategic planning cannot be theorized if its approaches and practices are neutral with respect to class, gender, age, race, and ethnicity (Sandercock 1998; Albrechts 2002). Strategic planning does not flow smoothly from one phase to the next; it is a dynamic and creative process. New points of view and facts that become known today might very well alter certain decisions made yesterday.

This "new" strategic spatial planning is presented not as a new ideology/ethics preaching a new world order but as a method for creating and steering a (range of) better future(s) for a place based on shared values (Ogilvy 2002). Its normative viewpoint produces quite a different picture from traditional planning in terms of: plans (strategic plans versus master plans or land use plans); type of planning (providing a framework and a justification for specific actions versus technical/legal regulation); type of governance (completely government-led versus government-led but negotiated form of governance); and content (visions and concrete actions that accept the full complexity of a place while focusing on local assets and networks in a global context, social-spatial quality, and a fair distribution of the pleasures and burdens). To some, the normative point of view may seem too broad a view of strategic spatial planning (Mintzberg 1994). However, the many European experiences documented in planning literature (Healey et al. 1997; Pascual and Esteve 1997; Albrechts et al. 2001; Albrechts et al. 2003; Pugliese and Spaziante 2003; Martinelli 2005a; Albrechts 2006; Motte 2006; Healey 2007; Balducci 2008) support this broader view.

Embedding transformative practices into strategic spatial planning

Transformative practices simply refuse to accept that the current way of doing things is necessarily the best way; they break free from concepts, structures, and ideas that persist only because of the process of continuity. It is precisely discontinuity that forces us outside the usual boundaries of "reasonableness" (De Bono 1992). Transformative practices focus on new concepts and new ways of thinking that change the way resources are used, (re)distributed, and allocated, and the way regulatory powers are exercised. They mobilize all necessary resources and develop the power to "travel" and "translate" into an array of practice arenas; and they transform these arenas, rather than merely being absorbed within them. Those concepts and ways of thinking that accumulate sufficient power to become customary may then "sediment" down into the cultural ground, which sustains ongoing processes and feeds into new strategic spatial processes (Hajer 1995; Albrechts and Lievois 2004; Healey 2005: 147–48; Healey 2006: 532). Transformative changes rarely occur overnight. Changes evolve in many small ways, building a basis of understanding and experiences that, over time, come together in what history may then describe as "a transformative moment" (Healey 2005: 158; Healey 2006: 541).

Focus on structural problems/issues

Transformative practices involve choice, valuation, judgment, and decisions that relate to human-determined ends and also to the selection of the most appropriate means of coping with and implementing such ends. This is contrary to practices as extensions of the present day. Transformative practices must be imagined as differing radically and structurally from the present reality, and result from judgments and choices formed with reference to the ideas of desirability and betterment. For spatial planners, both sustainable development and spatial quality generally provide lenses through which we can provide substance to desirability, and for the benefit of society. These are potentially rich concepts that may help to drive the policy integration of economic, environmental, and socio-cultural objectives in their spatial manifestation. Our concepts of sustainable development and spatial quality imply that a clear statement must be made against any notion of a purely quantitative approach to growth and in favor of the need for a "just" use of resources and social cohabitation (Hamilton 2004). If we look at plans today, most, if not all, embrace some unspecified notions of sustainability and spatial quality, although almost none of them questions growth as such.

The spectrum for transformative practices cannot be so open that the possibilities seem endless (Berger 1964; Ozbekhan 1969; Ogilvy 2002). Conditions and structural constraints on "what is" and "what is not" possible are placed by the past and the present. These conditions and constraints have to be questioned and challenged in the process, given the specific context of place and time. So in order to imagine the conditions and constraints differently, we need to deal with and overcome history. This defines the boundaries of a fairly large space between flexibility and inflexibility. In this way, transformative practices become the activity whereby (taking into account structural constraints) "that which might become" is "imposed" on "that which is," and it is "imposed" for the purpose of changing "what is" into "what might become." "Becoming" privileges change over persistence, activity over substance, process over product and novelty over continuity (Chia 2002: 866). This means a shift from an ontology of being, which privileges outcome and end-state, towards an ontology of becoming, in which actions, movement, relationships, process, and emergence are emphasized (Chia 1995: 601; 1999: 215). So this chapter argues that we should think in terms of the "heterogeneous becoming" of institutional transformation, the "otherness" of institutional outcomes, and the "immanent continuity" of institutional traces.

Change is the sum of many acts (individual, group, institutional) of re-perception and behavior change at every level. This takes decision-makers, planners, institutions, and citizens out of their comfort zones and compels them to confront key beliefs, to challenge conventional wisdom, and to look at the prospects of "breaking-out-of-the-box."

Planning requires a fine-grained analysis of what actually takes place in formal decision-making and implementation, in the transition from plan to formal adoption of the plan and in its actual implementation, as opposed to what most people normatively would like to see happen (Friedmann 1998). Research by Flyvbjerg (1998) makes it clear that critical analysis of cases is needed to discover the "whys and wherefores" of how elected representatives or preferential actors change the plan and the reasons why and methods by which executive officers depart from the formally approved plan.

From envisioning as a learning process to transformative practices

As futures are not just "out there" waiting to be discovered, we have to construct them. This is not a linear, but rather a dialectic (back casting and forecasting) process. Envisioning is the process by which individuals—or more preferably, groups—develop visions of future states for themselves, their organization, their city, or their region that are sufficiently clear and powerful to arouse and sustain the actions necessary for (parts of) these visions to become a reality (Goodstein et al. 1993). Visions themselves may not be seen as static descriptions of futures. They have to understand and portray the dynamic nature of development, changing challenges and contexts. We cannot confront complex dynamic realities with a language designed for simple static problems (Senge 1990)—hence the need for ways of thinking and for tools that help planners to cope with change in a dynamic environment (Winch 1998).

A vision must be placed within a specific context (economic, social, cultural, political, and power), place, time, and scale regarding specific issues that are of interest and within a particular combination of actors. The context provides the setting for the process, but it also takes form and undergoes changes in the process.

No one claims that the process of envisioning eliminates uncertainty with predictions; instead, it seeks to work with uncertainty as well as possible, and to enable actors to make decisions in view of the desired futures. According to Godet (2001: 8), envisioning is, above all, a state of mind (imagination and anticipation) that leads to behavior (hope and will). In the final analysis, we must come back to what "is" if we want to present ideas and concepts that are solid, workable, and of testable value. To arrive at these ideas, we need both the solidity of the analysis and the creativity of the design of alternative futures. To avoid naïve thinking, all of this must be rooted in an understanding of the basic processes and structural constraints that shape places (see Salet in this volume).

Since the envisioning of discontinuous futures involves change, all the usual forms of resistance to change (and especially structural change) are present. Since envisioning is so integral to the strategic planning process and so all-invasive, it cannot be confined to a single actor or institution in the process. We consider envisioning to be a collective process that concerns the future for which a wide variety of actors are personally responsible. Their vision, then, is more than a wish list—it involves commitment to the realization of the vision through practice (Friedmann 1987). A vision provides actors with views of the future that can be shared—a clear sense of direction, a mobilization of energy, and a sense of being engaged in something important (Goodstein et al. 1993). A vision is "communicatively rational" to the degree that it is reached consensually through deliberations involving relevant actors, where all are equally empowered and fully informed, and where the conditions of ideal speech are met (Innes 1996). The images provided in a vision involve a dynamic interaction between the relevant actors in the process rather than a unidirectional flow. The reiterative process occurs at the moment of creating the vision, as well as throughout the process of its implementation. The values and images of what a society wants to achieve are not generated in isolation but are socially constructed and given meaning and validated by the traditions of belief and practice; they are reviewed, reconstructed, and invented through collective experience (Ozbekhan 1969, but also Foucault 1980: 11; Hillier 1999; and Elchardus et al. 2000: 24).

Envisioning reveals how things can be different, how things could be truly better, how people can be innovative, how the natural creativity of the actors involved could be unlocked to improve our cities and regions, and how these natural tendencies that are typically inhibited or suppressed by the daily demands of our governance systems could be realized. The construction of different futures, which lies at the very heart of the transformative practices, requires creativity and original synthesis (Ozbekhan 1969: 87).

"Creative transformation" refers to changes in governance relating to current and historical relations of dominance and oppression (Young 1990). Planning and governance cannot be looked upon as separate, autonomous spheres within society, so in the following section, this chapter searches for a type of governance that will interlink with the transformative approach outlined above.

Governance[1]

Just as there are many traditions and collective practices, there are also many images of what regions, city regions, and cities want to achieve (Weeks 1993). The power constellation in a place determines what the problems and challenges of a place are and how they should/could be addressed. The opportunities for conceiving/implementing images are not equal. Some actors (individuals, groups, institutions) have more resources and power, which allow them to pursue their images. Therefore, power relations must be built into the conceptual framework of planning (Forester 1989; Sager 1994; Healey 1997a; Friedmann and Douglass 1998) and looked at in a given context of place, time, and scale regarding specific issues and particular combinations of actors.

Strategic spatial planning processes challenge established divisions of government and the cultures embedded in them. They also bring different models of governance and governance change face to face with one another (Healey 2006). This chapter argues that a feasible and efficient planning process should be centered on the elaboration of a mutually beneficial dialectic between top-down structural policies and bottom-up local uniqueness. Besides a bottom-up approach, rooted in conditions and potentialities of diversity (interpreted in their broadest sense), a complementary multilevel top-down policy aimed at introducing fundamental and structural changes is indispensable. Indeed, a mere top-down and centrally organized approach runs the danger of overshooting the local, historically evolved, and accumulated knowledge and qualification potential, while a one-dimensional emphasis on a bottom-up approach tends to deny—or at least to underestimate—the importance of linking local differences to structural macro-tendencies (Albrechts and Swyngedouw 1989). This dialectic constitutes the bare essence of multilevel governance.

Place policymaking is embedded in multiple institutional domains and interaction arenas. This blurs the meaning of traditional administrative boundaries and hierarchical settings in the development and implementation of policies (see the European INTERREG program). Initiatives to overcome fragmentation due to entrenched tiers of government and sectoral policy communities typically require a major institutional effort to achieve long-term effects (Albrechts et al. 2003). Moreover, the demand to transform the state in ways that will serve all relevant actors, especially the least powerful, and the emerging partnerships between governments and the private sector are provoking a shift towards more hybrid forms of democracy in a number of places in Europe.

Pluralist and inter-culturalist place

In Europe, some politicians, as well as planners, seem reluctant to involve the public in decision-making, because it involves giving up some control, and people who hold power are usually not inclined to give it up or share it. In other places, there is a tendency to involve major actors in the process. As spatial planning has almost no potential for concretizing strategies, it is necessary to involve relevant actors (public and private) needed for their substantive contribution, their procedural competences, and the role they might play in acceptance, in getting basic support and in providing (a kind of) legitimacy. However, planning has the potential to have an impact and to connect a very wide range of issues (from all kinds of actors with interests in a place to nature). These interests are potentially very diverse and conflicting. To overcome a commodified representation, nature must find a voice to reveal its intrinsic values (e.g. natural stability in ecosystems, biodiversity) as well as the more intangible cultural (e.g. aesthetic, symbolic) values (Hillier 1999; Sachs and Esteva 2003). Spatial planners capitalize on the "locus" using the characteristics of space and place, the natural as well as the built environment, the socio-spatial structure, the flows and the tissue (both spatial and social). In this sense, space gets its own relative autonomy. It serves as a medium and as an integration frame for human activities.

Some actors have the knowledge, skills, power, and networks through which they are able to influence or even steer planning proposals and policy decisions. Others lack the means and the cultural codes to participate in the system. Class, gender, race, and religion do matter in terms of whether citizens are included in the process (Young 1990). The creation of futures must be done under conditions of inequality and diversity. Any change has to deal with structural constraints, with issues of power and resistance, and with the irreconcilability of certain forms of interests. This requires a democratic polity that can encompass the realities of difference, inequality, and so on (Huxley 2000). The core is a democratic struggle: for inclusiveness in democratic procedures; for transparency in government transactions; for accountability of the state and planners to the citizens for whom they work; for the right of citizens to be heard and to have a creative input in matters affecting their interests and concerns at different scale levels; and for reducing or eliminating unequal power structures between social groups and classes (Friedmann and Douglass 1998). In several places in Europe, there is a call for the development of pluralist democratic tendencies in the wake of a crisis of representative democracy and a demand to transform the state in ways that will serve all of its citizens and especially the least powerful. Out of a shift towards a more hybrid democracy in some places, a type of governance has emerged that expands practical democratic deliberations rather than restricts them, that encourages diverse citizens' voices rather than stifles them, and that directs resources to basic needs rather than to narrow private gain. This type of approach uses public involvement to present real political opportunities, learning from action not only what works but also what matters. Through the involvement of citizens (and especially weak groups) in socially and politically relevant actions, some degree of empowerment, ownership, or acceptance is sought for these citizens (Friedmann 1992).

In Europe, increased personal mobility has made places more diversified. This can be seen either as a threat or as an opportunity. On the one hand, it can destabilize a place as migrants bring in habits, attitudes, and skills different from the original society. On the other hand, it can enrich and stimulate possibilities by creating hybrids, crossovers, and boundary blurring (Landry 2000: 264). Places must be creative with mutual understanding between cultures and ideas of equity (this is nothing less than a claim to full citizenship) (Sandercock 2003: 98). Inter-culturalism builds bridges, helps foster cohesion and conciliation, and produces new ideas out of the multicultural patchwork of places. This gives a voice to the minority groups or the otherwise socially excluded, so that their ideas are taken into account and brought into the process in order to change the planning, political decision-making, and implementation (Landry 2000).

Institutionalization

Government systems for development, control, and regulation have often been fixed for a long time; they are not fundamentally reviewed to adapt to changing circumstances. There are many examples to illustrate how difficult it is for an institution to change. The life of an institution often seems to be more important than what it does. Hence there is a need to view governance institutions not as a set of formal organizations and procedures established in law and "followed through" but rather as referring to the norms, standards, and morals of a society or social group that shape both the formal and informal ways of thinking and acting (Healey 2004a: 92). In some places, the process of "discourse structuration" and its subsequent "institutionalization" become perhaps more important than the plan as such (Hajer 1995; Albrechts 1999, 2003a, 2003b; Albrechts and Van den Broeck 2004). In this way, new discourses may become institutionalized and embedded in the norms, methods, attitudes, and practices, thus providing a basis for structural change. From there, a shared stock of values, knowledge, information, sensitivity, and mutual understanding may spread and travel through an array of regional, provincial, and local government arenas, sector departments, and consultants. Gradually, new approaches and new concepts can be sustainably embedded via institutionalization (Healey 1997a; Gualini 2001). Governments may call upon this intellectual capital when using its control function to reframe ways of thinking (Innes et al. 1994).

Multilevel governance

A multilevel governance approach would offer the potential to tease out causal linkages between global, national, regional, metropolitan, and local change, while also taking account of the highly diverse outcomes of such interactions. The dialectic between shifts in institutional sovereignty towards supranational regulatory systems (e.g. the possible impact of European directives for deregulation of public transport) and the principle of subsidiarity, which entails the rooting of policy action in local initiatives and abilities, illustrates the embeddedness of place policymaking in multiple institutional domains and interaction arenas, which blurs the meaning of hierarchical settings in the development of policies (Gualini 2001). Tensions may occur between the well-known scale and related government structure of a nested hierarchy from large to small or from top to bottom in terms of the reach of relationships in time and space (Healey 2004b; Albrechts and Lievois 2004).

In a new governance culture, the construction of arenas (who has to be involved, what is fixed and what is open in these arenas and which issues must be discussed), their timing (links to the strategic momentum), and the awareness that "fixed" may be a relative concept in some contexts all need careful reflection and full attention.

I linked envisioning to transformative practices, which involve constructing "desired" answers—images/visions and strategies for action—to the structural problems of our society. In Europe, there is a growing belief that strategic projects could provide a key to an action-oriented approach.

Strategic projects as a key to more "strategic" planning

Strategic projects are spatial projects, (preferably) coordinated by public actors in close cooperation with the private sector and other semi-public actors. These projects are strategic with respect to their capacity to achieve the visions, policy objectives, and goals embedded in strategic planning processes at different policy levels. They aim to transform the spatial, economic, and socio-cultural fabric of a larger area through a timely intervention. Strategic projects aim to integrate the visions, goals, and objectives from different policy sectors, as well as the ambitions and goals of the private sector. They also aim to integrate the inhabitants and users of the area. In this way, these projects may become transformative and integrative. They are strategic in the sense that they deal with specific key issues in an area. There are several possible types of projects (e.g. urban, rural, and new innovative employment). In line with the topic of this book, the chapter focuses on network-like projects.

Network-like projects

These projects are important within the European Spatial Development Perspective (CEC 1999). If we go back to the idea of a network, the main reason for actors (mega-city regions, institutions, investors) to enter into such a network is that it enables them to pursue a certain goal that is common to all of them and to achieve this goal in a better and easier way through cooperation rather than through acting individually (Albrechts and Mandelbaum 2005). To increase the likelihood that a network project will be realized, a close discourse is needed with those whose cooperation is necessary for its implementation (see Salet in this volume) and with those who are needed to win over the public (Flyvbjerg 1998). It is a dynamic endeavor that evolves in interaction both with the local, regional, national, and European contingencies and with external forces in order to address the agendas of those responsible for the implementation of a project and to build a broad basis of social support. Such a dialogue may enlarge the small "epistemic community" (Haas 1992; Faludi 1999) initially supporting the concept of a spatial network. This is definitely the case for many mega-city regions. Whatever can be done at the level of the network needs to involve progressively multilevel and inter-jurisdictional dimensions. Network concepts require the development of a complex governance setting throughout various institutional, political, territorial, and socio-economic contexts.

The European discourse on space is conceptualized in terms of the vocabulary of economic competition, with references to the underlying "growth rationale," but at the same time it is articulated with the aspiration of combining growth with concern for the environment. Network concepts articulate a spatial narrative based on a discursive process of "reimagining" territory and urban space within a country or across countries (see the INTERREG programs in Europe). In this discursive process of social spatialization, the driving rationale is one of competitiveness as the precondition for a possible wider acceptance of the concept.

Many of the European networks (in the making) (see Randstad, RhineRuhr, Flemish Diamond, etc.) are examples of polynucleated urban systems (see Dieleman and Faludi 1998; Hall in this volume). Their qualities are said not to be merely physical but also functional, in that the cities within the complex perform different functions.

The basic challenge in the process of (re)activating networks-in-the-making seems to be to find out what kind of strategic issues could attract the interest and sympathy of politicians, the business community, ordinary citizens, local communities, and municipalities, without these issues being presented in too antagonistic a manner, starting from the initial formulation of the issue and continuing on through the design of institutions and the arenas/forums needed to successfully conceive, plan, and implement the issue in a project. Such a project must reflect the specific politics (Healey 1997b) and context of the area (Zeng 2007). This context includes the history and geography. The selection of a strategic project is neither a neutral nor a random matter. From an analytical point of view, the strategic issues may be considered fairly obvious, but from a policy point of view they are definitely not. This phenomenon is common to most so-called urban networks (Dieleman and Faludi 1998). The spatial networks are dependent on a plethora of planning systems and planning control and building regulations, each with their own set of separate and exclusive objectives. This challenges the possibility of constructing a common discourse or a project within which conflicting objectives can be pursued equitably.

What can we learn from this? It is only too obvious that there is a wide gap between the socio-economic reality in polynuclear areas and the normative concept of spatial networks. It is abundantly clear that the concept of any spatial network will never automatically become part of the frame of reference for justifying decisions (Dieleman and Faludi 1998). In the first place, the concept must succeed in penetrating this frame of reference. Here there is stiff competition with other plans/projects and the (secret) agenda of a variety of actors (Faludi and Altes 1994). Moreover, it is obvious that the concept can be interpreted in different ways (Mandelbaum 1990).

The institutionalization of a network discourse is likely to prove difficult. The challenge is that of co-production, of the pursuit of joint results with other sectoral, social, and economic actors' activities and initiatives. In this respect, governance opens up a new potential field of collective action (Gualini 2001). Governance suggests an understanding of the linkage of different types of regulation in a territory in terms of political, social, institutional, and cultural integration and at the same time in terms of capacity to conceive and implement a vision or a specific project (Le Galès 1998). The need to design (informal) institutional arrangements for a network-like project that enmesh formal government structures and processes within the wider relational webs of economic and social life is becoming a key dimension of modern supra-local governance (see Healey 1997a: 290). This is especially the case for mega-city regions.

Epilogue

In Europe, planning is diverting from the idea of government as the sole provider of solutions to problems, and towards an idea of governance as the capacity to substantiate the search for creative and territorially differentiated solutions to problems, challenges, and opportunities. It is moving towards a more desirable scenario through the mobilization of a plurality of actors with different and even competing interests, goals, and strategies (Balducci and

Fareri 1999). The "new" strategic spatial planning presented in this chapter is conceived of as a democratic, open, selective, and dynamic process of co-production. It produces a vision that leads to a framework within which the problems and challenges can be understood, and provides a justification for short-term actions within a revised democratic tradition. A dissection of the process reveals the key elements that underlie this strategic planning: content and process; the static and the dynamic; constraint and aspiration; the cognitive and the collective; the planned and the learned; the socio-economic and the political; the public and the private; vision and action; the local and the global; legitimacy and a revised democratic tradition; values and facts; selectivity and integrativity; equality and power; the long term and the short term.

This chapter argues to apply the "lenses" of a reflective practitioner and of the (strategic) planning literature in an effort to broaden the concept and provide an alternative to address the structural challenges of our postmodern world in a constructive and progressive way. Strategic planning case studies (Healey et al. 1997; Pascual and Esteve 1997; Albrechts 1999; Albrechts et al. 2003; Pugliese and Spaziante 2003; Martinelli 2005a; Healey 2007; Balducci 2008) illustrate innovative practices. There is a need for inquiring into the epistemology of these practices, for making sense of what has been learned in action in relation to a wider context, and for testing the depth and comprehensiveness of these practices (Schön 1984). This should help efforts to evaluate and make sense of these practices in relation to a wider (theoretical) context. Abstract conceptualization and generalization of the accumulated knowledge of learning in action may help theorists to see some of what can be learned from practice. Strategic spatial planners, on the other hand, can be inspired and guided by new emerging theories.

The critical question of the leverage that the European strategic spatial planning exercises will achieve over time must be raised. Do they have the persuasive power to shift territorial development trajectories or—as some argue (Kunzmann 2001a, 2001b, 2001c)—are they little more than a cosmetic veil to hide the growing disparities evolving within Europe? A number of the European experiences (Healey et al. 1997; Pascual and Esteve 1997; CEC 1999; Albrechts et al. 2003; Albrechts et al. 2003; Pugliese and Spaziante 2003, Albrechts 2006; Motte 2006; Healey 2007; Balducci 2008) provide a fertile laboratory for advancing the understanding of the nature and potential of strategic spatial frameworks and strategies for twenty-first-century conditions.

Note

1 See for this chapter, and also for all other chapters, the bulk of material generated through the ESPON program (www.espon.eu).

References

Ackoff, R. (1981) *Creating the Corporate Future*, New York: John Wiley.

Albrechts, L. (1999) "Planners as Catalysts and Initiators of Change: The New Structure Plan for Flanders," *European Planning Studies*, 7(5): 587–603.

—— (2001) "In Pursuit of New Approaches to Strategic Spatial Planning," *International Planning Studies*, 6(3): 293–310.

—— (2002) "The Panning Community Reflects on Enhancing Public Involvement: Views from Academics and Reflective Practitioners," *Planning Theory and Practice*, 3(3): 31–47.

—— (2003a) "Planning and Power: Towards an Emancipatory Approach," *Environment and Planning C*, 21(6): 905–24.

—— (2003b) "Planning Versus Politics," *Planning Theory*, 2(3): 249–68.

—— (2004) "Strategic (Spatial) Planning Reexamined," *Environment and Planning B*, 31(5): 743–58.

—— (2006) "Shifts in Strategic Spatial Planning? Some Evidence from Europe and Australia," *Environment and Planning A*, 38(6): 1149–70.

Albrechts, L. and Lievois, G. (2004) "The Flemish Diamond: Urban Network in the Making," *European Planning Studies*, 12(3): 351–70.

Albrechts, L. and Mandelbaum, S. (eds.) (2005) *The Network Society: A New Context for Planning?*, New York: Routledge.

Albrechts, L. and Swyngedouw, E. (1989) "The Challenges for a Regional Policy Under a Flexible Regime of Accumulation," in L. Albrechts, F. Moulaert, P. Roberts, and E. Swyngedouw (eds.) *Regional Policy at the Crossroads*, London: Jessica Kingsley, pp. 67–89.

Albrechts, L. and Van den Broeck, J. (2004) "From Discourse to Facts: The Case of the ROM Project in Ghent, Belgium," *Town Planning Review*, 75(2): 127–50.

Albrechts, L., Alden, J., and Da Rosa Pires, A. (eds.) (2001) *The Changing Institutional Landscape of Planning*, Aldershot: Ashgate.

Albrechts, L., Healey, P., and Kunzmann, K. (2003) "Strategic Spatial Planning and Regional Governance in Europe," *Journal of the American Planning Association*, 69(2): 113–29.

Albrechts, L., Van den Broeck, J., Verachtert K., Leroy, P., and Van Tatenhove (1999) *Geïntegreerd Gebiedsgericht, Beleid: Een Methodiek* (*Integrated Area Policy: A Methodology*), Leuven/Nijmegen (in Dutch).

Amin, A. (2004) "Regions Unbound: Towards a New Politics of Place," *Geografisker Annaler*, 86B: 33–44.

Balducci, A. (2008) "Constructing (Spatial) Strategies in Complex Environments," in J. Van den Broeck, F. Moulaert, S. Oosterlynck (eds.) *Empowering the Planning Fields: Ethics, Creativity and Action*, Leuven: Acco, pp. 79–99.

Balducci, A. and Fareri, P. (1999) "Consensus-building, Urban Planning Policies, and the Problem of Scale: Examples from Italy," in F. Coenen, D. Huitema, and L. O'Toole (eds.) *Participation and the Quality of Environmental Decision Making*, Dordrecht: Kluwer Academic, pp. 163–78.

Barrett, S. and Fudge, C. (eds.) (1981) *Planning and Action*, London: Methuen.

Berger, G. (1964) *Phénoménologie du Temps et Prospective (Phenomenology of Time and Prospective)*, Paris: P.U.F. (in French).

CEC (1997) *The EU Compendium of Spatial Planning Systems and Policies*, Luxembourg: Office for the Official Publications of the European Communities.

—— (1999) *The European Spatial Development Perspective: Towards Balanced and Sustainable Development of the Territory of the EU*, Luxembourg: Office for the Official Publications of the European Communities.

Chapin, F.S. (1965) *Urban Land Use Planning*, Urbana, IL: University of Illinois Press.

Chia, R. (1995) "From Modern to Postmodern Organizational Analysis," *Organization Studies*, 16(4): 579–604.
—— (1999): "A 'Rhizomic' Model of Organizational Change and Transformation: Perspective from a Metaphysics of Change," *British Journal of Management*, 10(3): 209–27.
—— (2002) "Time, Duration and Simultaneity: Rethinking Process and Change in Organizational Analysis," *Organization Studies*, 22(6): 863–68.
Cullingworth, J.B. (1972) *Town and Country Planning in Britain*, London: Allen & Unwin.
De Bono, E. (1992) *Serious Creativity: Using the Power of Lateral Thinking to Create New Ideas*, New York: Harper Business.
Dieleman, F. and Faludi, A. (eds.) (1998) "Polynucleated Metropolitan Regions in Northwest Europe," *European Planning Studies*, Special Issue 6(4): 365–466.
Dyrberg, T.B. (1997) *The Circular Structure of Power*, London: Verso.
Elchardus, M., Hooghe, M., and Smits, W. (2000) "De Vormen van Middenveld Participatie" ("Forms of Bottom up Participation"), in M. Elchardus, L. Huyse, and M. Hooghe (eds.) *Het Maatschappelijk Middenveld in Vlaanderen* (*Bottom up Movements in Flanders*), Brussels: VUB Press, pp. 15–46 (in Dutch).
Faludi, A. (1999) *De architectuur van de Europese ruimtelijke ontwikkelingspolitiek (The Architecture of European Spatial Development Policy: Inaugural Lecture)*, Nijmegen: K.U. Nijmegen (in Dutch).
Faludi, A. and Altes, W. (1994) "Evaluating Communicative Planning: A Revised Design for Performance Research," *European Planning Studies*, 2(4): 403–18.
Flyvbjerg, B. (1998) *Rationality and Power Democracy in Practice*, Chicago, IL: University of Chicago Press.
Forester, J. (1989) *Planning in the Face of Power*, Berkeley, CA: University of California Press.
Foucault, M. (1980) *The History of Sexuality*, New York: Vintage.
Friedmann, J. (1982) *The Good Society*, Cambridge, MA: MIT Press.
—— (1987) *Planning in the Public Domain: From Knowledge to Action*, Princeton, NJ: Princeton University Press.
—— (1992) *Empowerment: the Politics of Alternative Development*, Oxford: Blackwell.
—— (1998) "Planning Theory Revisited," *European Planning Studies*, 6(3): 245–53.
Friedmann, J. and Douglass, M. (1998) "Editor's Introduction," in M. Douglass and J. Friedmann (eds.) *Cities for Citizens*, Chichester: John Wiley & Sons, pp. 1–6.
Godet, M. (2001) *Creating Futures: Scenario Planning as a. Strategic Management Tool*, London: Economica.
Goodstein, L., Nolan, T., and Pfeiffer, J. (1993) *Applied Strategic Planning*, New York: McGraw-Hill.
Gualini, E. (2001) *Planning and the Intelligence of Institutions*, Aldershot: Ashgate.
Haas, P. (1992) "Introduction: Epistemic Communities and International Policy Coordination," *International Organization*, 46(1): 1–35.
Hajer, M. (1995) *The Politics of Environmental Discourse*, Oxford: Oxford University Press.
Hamilton, C. (2004) *Growth Fetish*, London: Pluto Press.
Healey, P. (1997a) *Collaborative Planning, Shaping Places in Fragmented Societies*, London: Macmillan.

—— (1997b) "The Revival of Strategic Spatial Planning in Europe," in P. Healey, A. Khakee, A. Motte, and B. Needham (eds.) *Making Strategic Spatial Plans*, London: U.C.L. Press, pp. 3–19.

—— (2004a) "Creativity and Urban Governance," *Policy Studies*, 25(2): 87–102.

—— (2004b) "The Treatment of Space and Place in the New Strategic Spatial Planning in Europe," *International Journal of Urban and Regional Research*, 28(1): 45–67.

—— (2005) "Network Complexity and the Imaginative Power of Strategic Spatial Planning," in L. Albrechts and S. Mandelbaum (eds.) *The Network Society: A New Context for Planning?* New York: Routledge, pp. 146–60.

—— (2006) "Relational Complexity and the Imaginative Power of Strategic Spatial Planning," *European Planning Studies*, 14(4): 525–46.

—— (2007) *Urban Complexity and Spatial Strategies*, London: Routledge.

Healey, P., Khakee, A., Motte, A., and Needham, B. (eds.) (1997) *Making Strategic Spatial Plans*, London: University College London Press.

Hillier, J. (1999) "What Values? Whose Values?" *Ethics, Place and Environment*, 2(2): 179–99.

—— (2005) "Straddling the Post-structural Abyss: Between Transcendence and Immanence?" *Planning Theory*, 4(3): 271–99.

—— (2007) *Stretching Beyond the Horizon*, Aldershot: Ashgate

Huxley, M. (2000) "The Limits of Communicative Planning," *Journal of Planning Education and Research*, 19(4): 369–77.

Innes, J. (1996) "Planning through Consensus-building: A New View of the Comprehensive Planning Ideal," *Journal of the American Institute of Planners*, 62 (4): 460–72.

Innes, J., Grüber, J., Thompson, R., and Neuman, M. (1994) *Coordinating Growth Management through Consensus Building: Incentives and the Generation of Social, Intellectual and Political Capital*, Berkeley, CA: University of California at Berkeley, Institute of Urban and Regional Development.

Kreukels, A. (2000) "An Institutional Analysis of Strategic Spatial Planning: the Case of Federal Urban Policies in Germany," in W. Salet and A. Faludi (eds.) *The Revival of Strategic Spatial Planning*, Amsterdam: Royal Netherlands Academy of Arts and Sciences, pp. 53–65.

Kunzmann, K. (2001a) "State Planning: a German Success Story?" *International Planning Studies*, 6(2): 153–66.

(2001b) "L'aménagement du Territoire en Allemagne: Une Discipline sans Utopias?" ("Spatial Planning in Germany: A Discipline without Utopias?"), *Territoires 2020*, 4(2001): 101–16.

—— (2001c) "The Ruhr in Germany: A Laboratory for Regional Governance," in L. Albrechts, J. Alden, and A. Da Rosa Pires (eds.) *The Changing Institutional Landscape of Planning*, Aldershot: Ashgate, pp. 133–58.

Landry, C. (2000) *The Creative City: A Toolkit for Urban Innovators*, London: Earthscan.

Le Galès, P. (1998) "Regulations and Governance in European Cities," *International Journal of Urban and Regional Research*, 22(3): 482–506.

Mandelbaum, S.J. (1990) "Reading Plans," *Journal of the American Planning Association*, 56(3): 350–56.

Martinelli, F. (ed.) (2005a) *La Pianificazione Strategica in Italia e in Europa: Methodologie ed esiti a Confronto* (*Strategic Planning in Italy and in Europe: Methodology*), Milano: Franco Angeli (in Italian).

—— (2005b) "Introduzione" ("Introduction"), in F. Martinelli (ed.) *La Pianificazione Strategica in Italia e in Europa: Methodologie ed esiti a Confronto* (*Strategic Planning in Italy and in Europe: Methodology*), Milano: Franco Angeli, pp. 11–31 (in Italian).

Masboungi, A. and De Gravelaine, F. (2002) *Projets Urbains en France (Urban Projects in France)*, Paris: Editions du Moniteur (in French).

Mastop, H. and Faludi, A. (1997) "Evaluation of Strategic Plans: the Performance Principle," *Environment and Planning B*, 24 (6): 815–22.

Mintzberg, H. (1994) *The Rise and Fall of Strategic Spatial Planning*, New York: The Free Press.

Motte, A. (1994) "Innovation in Development Plan Making in France 1967–93," in P. Healey (ed.) *Working Paper 42, Newcastle upon Tyne, Department of Town and Country Planning*, Newcastle: University of Newcastle upon Tyne, pp. 90–103.

—— (2006*): La notion de Planification Stratégique Spatialisée en Europe (1995–2005)* (*The Concept of Strategic Spatial Planning in Europe (1995–2005)*), Lyon: Puca (in French).

Ogilvy, J. (2002) *Creating Better Futures*, Oxford: Oxford University Press.

Ozbekhan, H. (1969) "Towards a General Theory of Planning," in E. Jantsch (ed.) *Perspective of Planning*, Paris: OECD, pp. 45–155.

Pascual, I. and Esteve, J. (1997) *La Estrategia de Las Ciudades. Planes Estratégicos Como Instrumento: Métodos, Téchnicias y Buenas Practices* (*The Strategy of Cities: Strategic Plans as Instrument :Methods, Techniques and Good Practices*), Barcelona: Diputacion de Barcelona (in Spanish).

Perloff, H. (1980) *Planning and the Post-Industrial City*, Washington, DC: Planners Press.

Pugliese, T. and Spaziante, A. (eds.) (2003) *Pianificazione Strategica per le Città: Riflessioni Dale Pratiche (Strategic Planning for the City: Reflections from Practice)*, Milano: F. Angeli (in Italian).

Sachs, W. and Esteva, G. (2003) *Des ruines du Développement* (*The Ruins of Development*), Paris: Le Serpent à Plumes (in French).

Sager, T. (1994) *Communicative Planning Theory*, Aldershot: Avebury.

Salet, W. and Faludi, A. (eds.) (2000) *The Revival of Strategic Spatial Planning*, Amsterdam: Royal Netherlands Academy of Arts and Sciences, pp. 1–10.

Sandercock, L. (1998) *Towards Cosmopolis: Planning for Multicultural Cities*, Chichester: John Wiley and Sons.

—— (2003) *Cosmopolis ll. Mongrel Cities in the Twenty-First Century*, London: Continuum.

Scharpf, F.W. and Schnabel, F. (1978) "Durchsetzungsprobleme der Raunordnung im öffentlichen Sektor" ("Problems of Implemention in Spatial Planning"), *Informationen zur Raumentwicklung*, 1: 19–28 (in German).

Schön, D. (1984*) The Reflective Practitioner: How Professionals Think in Action*, Aldershot: Avebury.

Secchi, B. (1986) "Una Nuova Forma di Piano" ("A New Form of Plans"), *Urbanistica*, 82: 6–13 (in Italian).

Senge, P. (1990) *The Fifth Dimension*, New York: Doubleday.

Van den Broeck, J. (1987) "Structuurplanning in Praktijk: Werken op Drie Sporen" ("Structure Planning in Practice: Working on Three Tracks"), *Ruimtelijke Planning ll.A.2.C* : 53–119 (in Dutch).

Weeks, J. (1993) "Rediscovering Values," in J. Squires (ed.) *Principled Positions: Postmodernism and the Rediscovery of Value*, London: Lawrence & Wishart, pp. 189–211.

Winch, G. (1998) "Dynamic Visioning for Dynamic Environments," *The Journal of the Operational Research Society*, 50(4): 354–61.

Young, I. (1990) *Justice and the Politics of Difference*, Princeton, NJ: Princeton University Press.

Zeng, X. (2007) *Strategic Planning for a Balanced Urban Development: Case Study of the Commercial Property Development in Pearl River New Town, Guangzhou, China*, HS Program Master's Thesis, University of Leuven, Belgium.

Part II

Multi-polity governance and planning in federacy

5 Novel spatial formats

Megaregions and global cities

Saskia Sassen

Introduction

Major shifts in the scales, spaces, and contents of economic activity are engendering novel spatial formats. The more prominent of these include global cities and megaregions, both of which are contributing to a whole series of old and new global intercity geographies. In turn, these shifts call for changes in our interpretations and policy frameworks to adjust to these novel spatial formats and maximize their benefits and distributive potential. Rather than reviewing all the components of this transformation, I will examine the spatial features and economic contents of megaregions and global cities. The policy issues are sufficiently complex and diverse across administrative units to warrant a separate discussion; they are not addressed here.[1]

While megaregions and global cities are different formats, I will argue that analytically we can identify similar dynamics at work in both arrangements. Two such dynamics stand out. One is scaling and its consequences—in this case, megaregional scaling and global scaling. The other is the interaction between geographic dispersal and new kinds of agglomeration economies, which, in this case, are operating respectively, within a megaregion and in global cities. Specifying a common analytic ground for these two very diverse spatial formats should enable us to develop a sharper approach to empirical research and, possibly, policy. These diverse spatial formats should also help in assessing the extent to which policy decisions can encourage greater economic integration between a country's more globalized city (or cities) and other areas currently performing subordinate functions within the national territorial hierarchy. In other words, taking a megaregional scale might help in connecting the "winners" and the "laggards." The megaregion then becomes a scale that includes both globalizing and provincial cities and areas. This raises a question as to whether this linking of winners and laggards can also be extended to cross-border intercity networks by strengthening the connections between winners and laggards in the global political economy.

One consequence is that not only would winners gain privileges, as is typical with the "targeting" of resources to enable the formation of world-class cities and silicon valleys, but also laggards would have a chance to do better.

More precisely, laggards can be enabled to become dynamically interconnected with winners within a megaregion in ways that replicate current practices at the global scale, notably outsourcing to low-cost areas. But they would do so in novel ways made possible because the low-cost area and the high income area are within a megaregion. The hope is that rather than pursuing the usual economic policies focused on the most advanced sectors, this would make a strong case for concentrating on the poorer regions not as charity but as a recognition that they are part of the advanced sectors; after all, when major firms outsource jobs to low-cost areas across the world, they are outsourcing some of *their* tasks. Many advanced economic sectors sufficiently combine diverse tasks so that they have preference for both lower-cost areas for some of these tasks, and yet also for dense high-cost areas for other tasks.

To mention just one of several possible examples, this type of framing would bring value to poorer areas within even the most developed countries, as these might be developed to house activities that are now outsourced to low-wage countries. One key focus should be to avoid a race to the bottom, which occurs when these activities are offshored; it would be simpler to avoid this race to the bottom when both headquarters and low-wage activities are in the same country. A second goal should be to provide alternative or complementary development paths to what is today's prevalent path, i.e. the preference for policies to support high-end economic activities, such as bio-tech parks and luxury office parks.

Parallel to this effort to incorporate laggards into policy frames that currently target mostly successful areas is the effort to understand how cities in the middle range of urban hierarchies fit in today's global intercity geographies. For instance, in the case of the US, many of these mid-range cities are also part of megaregions. The analytic bridge between megaregions and intercity geographies is the fact that the operational chains of a growing number of firms are currently part of both spatial formats. This opens up a whole new research agenda about economic globalization and place, in addition to the existing global city scholarship. Here, I will examine 2008 data from a recent study of the seventy-plus global hubs, which include nine US cities. The purpose here is to broaden the range of globalizing cities in the analysis beyond the usual top twenty global cities, and to weaken the overwhelming analytical and empirical weight of these major cities in our understanding of cities in a global political economy. Similarly, examining the top nine US cities is a way to expand the analysis beyond the familiar top three or four global cities in the US. It shows that as corporate economic globalization has developed so has the quantity and types of territorial insertions of global city functions; in the case of the US, these presently go well beyond just New York City. This expanded focus also allows us to capture the growth and diversification of global intercity flows.

These are some of the issues addressed in this chapter. The first section is a brief introduction to megaregions and global cities, as well as their intercity geographies. The second section brings together critical analytic issues for the study of these two spatial formats. The third and fourth section then apply these to megaregions and the proliferation of major and minor global cities, respectively.

Two novel spatial formats

One major tendency evident throughout the world is the formation of increasingly large urbanized areas that, at some point, are best described as megaregions. Here, I take the Regional Plan Association (RPA) definition of ten emergent megaregions in the US, each at different levels of development (Goldfeld 2007). Often, these are merely seen as more of the same—more people, more endless urban landscapes. At its most elemental stage, the megaregion is born from population growth in a geographic setting where cities and metro-areas blend into each other. This indeed calls for cross-regional infrastructures, notably transport and electricity, and various forms of regional planning and coordination, which can be seen at the present time. But are these conditions, which amount to an expanded version of urbanization economies, all there is to consider?

A basic starting point for my analysis is that a megaregion is a sufficient, internally diverse, economic territory that can contain diverse spatial logics—particularly, agglomeration and dispersal logics—which might translate into high-cost high-density areas and low-cost low-density areas. We know that large, integrated firms need both types of areas for their operations. Thus, the megaregional scale could enable the exploring of novel development strategies predicated on this diversity of spatial logics, hopefully to the advantage of both the more advanced and the least advanced areas within that megaregion. It would take innovative governance umbrellas, along with new types of private-public arrangement.

The most common advantages of scales larger than that of the city, such as metropolitan and regional scales, come from sharing transport infrastructures for people and goods, enabling robust housing markets, and, possibly, supporting the development of office, science, and technology parks. More complex and elusive is whether the benefits of megaregional economic interaction can go beyond these familiar scale economies. There is no definitive research on this subject. Thus empirical specification can be only partial, as the available evidence is fragmentary for the urban level, a shortcoming that becomes acute when dealing with the novel category of the megaregion.[2] There is, however, enough analysis and evidence on one particular component of this subject—the advantages for global firms and markets of particular types of agglomeration economies at the urban level—that we can begin to use it as a lens into the megaregional scale that can take us beyond merely urbanization economies. Agglomeration economies are to be distinguished from such familiar urbanization economies because they involve complex interactions of diverse components, rather than simply, for instance, more people using a train line and the scale economies this might enable.

Departing from the more common propositions, I will argue that the specific advantages of the megaregional scale consist of and arise from the co-existence within one regional space of multiple types of agglomeration economies. These types of agglomeration economies are currently distributed across diverse economic spaces and geographic scales: central business districts, office parks, science parks, the transportation and housing efficiencies derived from large (but not too large) commuter belts, low-cost manufacturing districts (nowadays often offshore), tourism destinations, specialized branches of agriculture, such as horticulture or organically grown food, and the complex kinds of agglomeration economies evident in global cities. Each of these spaces evinces distinct agglomeration economies and empirically, at least, is found in diverse types of geographic settings—from urban to rural, from local to global.

The thesis is that a megaregion is sufficiently large and diverse, and has the capability to accommodate a far broader range of types of agglomeration economies and geographic settings than it typically does today. This would take the advantages of megaregional location beyond the notion of urbanization economies. A megaregion can then be seen as a scale that can benefit from the fact that our complex economies need diverse types of agglomeration economies and geographic settings. This diversity ranges from extremely high agglomeration economies shown by the specialized advanced corporate services to the fairly modest economies demonstrated by suburban office parks and regional labor-intensive low-wage manufacturing. The megaregion can incorporate this diversity into a single economic mega-zone. Indeed, in principle it could create conditions for the return of particular (not all) activities now outsourced to other regions or to foreign locations.[3]

Thus, the critical dimension for the purposes of this chapter is not merely a question of the contents of a megaregion, such as its economic sectors, transport infrastructure, housing markets, and the types of goods and services that get produced and distributed, exported and imported—a sort of X-ray of a megaregion. Another critical aspect is the specification of economic interactions within the megaregion in order to detect what could be reincorporated into that region (e.g. factories, or routine clerical work that is now outsourced, nationally or internationally) as well as to detect emerging megaregional advantages.

One path into the question of megaregions and the global economy is through the particular global circuits that provide connections across borders. The diversity of sub-economies in a megaregion signals a possibly equally diverse set of such cross-border circuits (Sassen 2008a, 2008c; SAIS Review 2009). Identifying these circuits is significantly easier for some of these sub-economies (e.g. Silicon Valley, Hollywood) than for others (e.g. the Midwestern machine and tool-making export-oriented sector, Wall Street). The best data available today on these global circuits, beyond specialized firm studies and commodity-chains (e.g. Wal-Mart) concern global financial and specialized services, which are mostly intercity networks. Particular networks connect particular groups of cities. This allows us to recover multiple details about the diverse roles of cities in the global economy.

The formation of intercity geographies is contributing a socio-technical infrastructure for a new global political economy, new cultural spaces, and new types of social networks. Some of these intercity geographies are thick and highly visible–the flows of professionals, tourists, artists, and migrants among specific groups of cities. Others are thin and barely visible—the highly specialized electronic financial trading networks that connect particular cities depending on the type of instrument involved. A bit thicker are the global commodity chains for diverse products that run from exporting hubs to importing hubs. Later on in this chapter, I will briefly examine some of these specialized intercity geographies.

An often overlooked dimension underlying these intercity geographies, and one that I keep stumbling upon in my research, is that today's global economy brings to the fore the specialized capabilities of different cities and regions. This goes against the more common notion that globalization homogenizes urban economies, a notion that I argue is only partially correct. Globalization homogenizes standards (for manufacturing, for the building of state-of-the-art office districts, for financial reporting, for accounting, and so on), and it engenders global markets for standardized products. It also feeds the specialized differences of places—thus, Chicago and New York, the two major financial centers of the US, have each become more and more specialized in their distinct sectors. The same can be said for Shanghai, Hong Kong, and Shenzhen, the three major financial centers of China—they are not becoming more equal. This contributes to explain why the number of major and minor global cities has expanded parallel to the expansion of globalization. One effect of these trends is to multiply the number of specialized/distinct circuits connecting cities around particular economic activities.

The specialized economic histories of major cities and regions matter in today's global economy because there is a globally networked division of functions. This fact is easily obscured by the common emphasis on competition and by the standardization (no matter how good the architecture) of state-of-the-art built environments, from offices to airports. This, then, also means that today's megaregions need to extract these specialized capabilities, which might include some very different sub-economies in different sites of a given megaregion. It is important to standardize transport infrastructures and various standards across a megaregion. However, this should not obscure the fact that the value-adding potential of that region may well lie in the particular economic (and cultural) capabilities of the diverse urban and non-urban sites of that region. These types of particularities come to the fore in the evidence on the different types of advantages of diverse cities worldwide. There is no perfect global city. Global economic actors (but also cultural, political, and civic actors) need many global cities, no matter how imperfect, rather than just one perfect global city.

This also means that a city's or a region's role in these intercity geographies is not only determined by its overall rank—an aggregate measure—but in fact can be critically shaped by its specialized capabilities. Elsewhere (Sassen 2008a), I have argued that the common notion of the homogenization of the urban landscape in today's economy misses a critical point. It neglects, or obscures, the fact of the diversity of economic trajectories through which cities and regions become globalized, even when the final visual outcomes may look similar. Out of this surface analysis based on homogenized landscapes comes a second possibly spurious inference: that similar visual landscapes are a function of convergence. Both propositions—that similar visual landscapes are indicators both of similar economic dynamics and of convergence—may indeed capture various situations. But key conditions are not captured, and, in fact, are rendered invisible by such notions. Similar landscapes may contain very different economies and hence may not be competing—they are, instead, complementary. At the scale of the megaregion, this can become very significant because it signals that a megaregion can accommodate a broad range of a large firm's diverse activities, as I discuss in the next section.

The diverse spatial logics of megaregions: does their interaction add to economic growth?

Translocal chains of operations are increasingly common for many firms and for whole economic sectors.[4] Establishing which segments of such chains are located in a given area becomes important in the specification of that area. One set of familiar categories through which to establish this is the agglomeration of a firm's operations versus its dispersal.

My thesis is that the megaregion is a scale that can accommodate a very broad range of the segments in a firm's or a sector's chain of operations (from those subject to agglomeration economies to those that can be dispersed to low-cost, low-density areas). Such a firm or sector can then derive multiplier growth effects from the fact that the region can accommodate both the agglomeration and the dispersal components of its operational chain. Thereby this thesis captures one very specific and partial feature of megaregions, but one overlooked in the emergent discussion about such regions. By emphasizing this particular dynamic, I do not want to minimize the significance of the more familiar facts of urbanization economies and scale economies. Both are of importance, and can make an enormous difference to the economic well-being of a region. I merely want to add yet another variable and argue that part of the specificity of the scale of the megaregion is that it *can* accommodate such diverse, yet mutually dependent spatial logics in ways that would be almost impossible in cities or suburbs. Today, these dependencies are largely organized at a global scale, with global outsourcing the most familiar format. The substantive rationality for the pursuit of my thesis is that it contains the possibility of significant equity effects that are lost in global outsourcing—

where it often becomes a race to the bottom regarding workers' wages and work conditions.

Thus, the first step is to examine why and at what stage of the production process a firm's or a sector's agglomeration sites are articulated with its outsourcing, or, more generally, its geographic dispersal sites. Examining the question of agglomeration economies in the current period is framed by two facts that are potentially in tension with each other. On the one hand, the new information technologies enable firms to disperse a growing range of their operations, whether at the metro, regional, or global level, without losing system integration.[5] This has the potential to reduce (though not eliminate) the benefits of urbanization economies for such firms. Whether this also reduces agglomeration economies for these firms is, in my analysis, a more complicated question. In my research, I find that for some headquarters, such a move brings with it a greater need to buy inputs that used to be produced in-house, and that it is these inputs whose production continues to evince agglomeration economies, and hence tends to be bought from firms located in cities (Sassen 2001: Chapter 5). On the other hand, the evidence clearly shows a preference for high-density locations for firms in a large range of advanced economic sectors; such locational preferences point to the existence of agglomeration economies, although the character of the latter has also changed (Sassen 2001: Chapters 1 and 5).

First, I address the most extreme instance—globalized firms with considerable digitization of their production process and their outputs. In this case, there are conceivably fewer and fewer agglomeration advantages, especially for the most advanced sectors, typically high-value producing, capable of buying the latest technologies, and highly globalized, that is, with multiple operations across the world. Contesting this technologically driven explanation, one can show how and why, precisely, these firms are subject to extreme agglomeration economies in some—although not all, of their components.[6] This fact matters for understanding megaregional advantage because megaregions contain extremely dense cities with diverse resources and types of talent that can deliver the agglomeration economies that leading firms need. A second implication for the megaregion is that it can accommodate low-cost, underdeveloped areas—that is, the second type of sites that characterize so many of the leading economic sectors.

What gives this additional meaning for megaregions are the growth effects resulting from interactions of a firm's diverse types of sites. Thus, one of the hypotheses in my global city model is that a firm's central headquarter functions expand as a result of that firm's multi-sitedness (whether national or global).[7] The more affiliates, service outlets, or factories a firm has worldwide, the more its central management and servicing functions expand. This is a significant growth potential that can be endogenized into a megaregion. This growth effect goes beyond the mere addition of jobs resulting from that of a megaregion capturing more sites from a firm's chain of operations, in itself an attractive potential. It also adds a complex networked dynamic that begins to articulate very diverse parts of a megaregion.

Does geographic dispersal feed agglomeration economies?

A good starting point is to focus on why the most advanced firms of the knowledge economy are subject to what are often seen as extreme agglomeration economies, even when they function in electronic markets and produce digitized outputs. Another way to ask this question is to focus on the most globalized and digitized of all knowledge sectors: Why does global finance need financial centers? Or, more generally, why do highly specialized global corporate services that can be transmitted digitally thrive in dense downtowns? This means inserting place in an analysis of knowledge economies, which are usually examined in terms of their mobility and space-time compression. Looking at the knowledge economy, and more broadly, global firms, from the optic of regions, cities, or metro-areas, brings in different variables.[8]

In its sharpest formulation, the question then becomes: Why do cities matter for global firms, particularly global firms that are wealthy enough to buy the technical innovations that free them from place, its frictions, and its costs? Further, why do they matter in a way that they did not as recently as the 1970s, when major headquarters were far more likely to leave cities and relocate in suburbs. Here are three logics that contribute to an answer.[9]

The first is that no matter how intensive a user of digital technology a firm is, its operational logic is not the same as the engineer's logic for designing that technology. Confusing these two potentially very diverse logics has produced a whole series of misunderstandings. When the new information and communications technologies (ICTs) began to be widely used in the 1980s, many experts "forecasted" the end of cities as strategic spaces for firms in advanced sectors. Many routinized sectors did leave cities, and many firms dispersed their more routine operations to the regional, national, and global scale. But the most advanced sectors and firms kept expanding their top-level operations in particular types of cities.

Why were those experts so wrong? They overlooked a key factor—when firms and markets use these new technologies, they do so with financial or economic objectives in mind, not the objectives of the engineer who designed the technology. The logics of users may well thwart or reduce the full technical capacities of the technology.[10] When firms and markets disperse many of their operations globally, with the help of the new technologies, the intention is not to relinquish control over these operations but rather to keep control over top-level matters and to be capable of appropriating the benefits/profits of that dispersal.[11] Insofar as central control is part of the globalizing of activities, their top-level headquarter functions have actually expanded simply because it is more complicated and riskier to function in 30 or 50 or more countries, each with distinct laws, accounting rules, and business cultures.

As these technologies are increasingly helpful in maintaining centralized control over globally dispersed operations, their use has also fed the expansion of central operations. The result has been an increase in high-level office operations in major cities, and a growth in the demand for high-level and highly paid professional services, either produced in-house or bought from specialized service firms. Thus, the more these technologies enable global geographic dispersal of corporate activities, the more they produce density and centrality at the other end—the cities where their headquarter functions are located.

A second logic explaining the ongoing advantages of spatial agglomeration has to do precisely with the complexity and specialization level of central functions. These rise with globalization and with the added speed that the new ICTs allow for. As a result, global firms increasingly need to buy the most specialized financial, legal, accounting, consulting, and other such services. These service firms get to do some of the most difficult and speculative work. It has been increasingly found that these corporate service firms evince agglomeration economies, as their work benefits from being in complex environments that function as knowledge centers, because they contain multiple specialized firms and high-level professionals with worldwide experience. Cities are such environments—with the forty-plus global cities in the world the most significant of these environments, but with a growing number of other cities strong in particular elements of such environments. In brief, cities or central places provide the social connectivity that allows a firm to maximize the benefits of its technological connectivity.[12]

A third logic concerns the meaning of information in an information economy, of which there are two types. One is the datum, which may be complex, and yet is standard knowledge—the level at which a stock market closes, the privatization of a public utility, a bankruptcy. But there is a far more difficult type of "information," akin to an interpretation/evaluation/judgment. It entails negotiating a series of data, and a sequence of interpretations of a mix of data in hopes of producing a higher-order datum. Access to the first type of information is now global and immediate (even if often at a premium) from just about any place in the highly developed world, and increasingly in the rest of the world thanks to the digital revolution.

It is the *second* type of information that requires a complicated mixture of elements—the "social infrastructure" for global connectivity—which gives major financial centers a leading edge. When the more complex forms of information needed to execute major international deals cannot be gotten from existing databases, no matter how much one is willing to pay, then one needs to create that information; it becomes part of the production process in specialized corporate service firms, including financial services both as service providers and as firms in their own right. That creation includes, as critical components, interpretation, inference, and speculation. At this point, one needs the social information loop and the associated de facto interpretations and inferences that come with bouncing off information among talented, informed people. It is the importance of this input that has given a whole new significance to credit rating agencies, for instance. Part of the rating has to do with interpreting and inferring. When this interpreting becomes "authoritative," it becomes "information" available to all. For specialized firms in these complex domains, credit ratings are but one of these inputs; the creation of authoritative information needs to be part of a production process, either in-house or bought from specialized firms. This process of making inferences/interpretations into "information" takes an exceptional mix of talents and resources. Cities are complex environments that can deliver this mix.

The key implication of this analysis for megaregions is the possibility of containing at least some of the dispersed operations of a given firm, as well as its central headquarters. The feedback effects of containing both of these can be significant, feeding simultaneous growth in a megaregion's low-cost, and possibly marginal areas, as well as in its global cities, or cities that are national business centers.

Cities in the global economy: an expanding network with regional groundings

A new data set shows nine US cities, rather than the usual four, to be major commercial hubs in the global economy. In addition to New York, Chicago, Los Angeles, and Boston, the list now includes San Francisco, Atlanta, Miami, Houston, and Washington, DC.

The data were collected from the perspective of firms and investors who want to do business globally. The study gets at equity issues through variables such as basic services and quality of life, which brings down some of the major centers, which are characterized by large numbers of poor and degraded residential and commercial areas. But the study does not really address issues of inequality per se, which we know from other sources are critical in global cities (Sassen 2001: Chapters 8 and 9; Sassen 2008c; Annals of the American Academy of Political and Social Science 2009).

The data come from the MasterCard Global Centers of Commerce (GCOC) study, which shows how major cities compare in performing critical functions that connect markets and commerce globally, and was developed by a panel of eight experts from different parts of the world.[13] In the 2006 phase, we started out with sixty-three cities that met the initial criteria; in the 2008 phase, this number had increased to seventy-five, which was eventually aggregated into seven overall indicators (legal and political framework, economic stability, ease of doing business, financial flow, business center, knowledge creation/information flow, livability). The result was that seventy-five cities qualified as major and not-so-major commercial hubs in the global economy, including nine US cities (see Table 5.1 and Table 5.2 (parts 1, 2, and 3).

Several major trends stand out. One of the most interesting findings coming out of the 2008 MasterCard study is a new type of regionalism (see Table 5.1). Even as the network of major and minor global cities expands, we see a

Table 5.1 Top thirty cities for selected indicators (2008)

Rank	*GCOC Index**	*Indicator 1: Political and legal frameworks*	*Indicator 3: Ease of doing business*	*Indicator 5: Business center*
1	London	Stockholm	Singapore	Hong Kong
2	**New York**	Singapore	Hong Kong	London
3	Tokyo	Copenhagen	London	Singapore
4	Singapore	**New York**	Toronto	Shanghai
5	**Chicago**	**Chicago**	**New York**	Dubai
6	Hong Kong	**Philadelphia**	Dublin	Tokyo
7	Paris	**Los Angeles**	Edinburgh	Paris
8	Frankfurt	**Boston**	Vancouver	**New York**
9	Seoul	**Atlanta**	Montreal	Amsterdam
10	Amsterdam	**Miami**	**Chicago**	Seoul
11	Madrid	**San Francisco**	**San Francisco**	Frankfurt
12	Sydney	**Houston**	Sydney	**Los Angeles**
13	Toronto	**Dallas**	**Los Angeles**	Bangkok
14	Copenhagen	**Washington, DC**	**Boston**	**Chicago**
15	Zürich	Zürich	**Washington, DC**	**Miami**
16	Stockholm	Geneva	Copenhagen	Taipei
17	**Los Angeles**	Toronto	**Atlanta**	Madrid
18	**Philadelphia**	Montreal	**Miami**	Milan
19	Osaka	Vancouver	Melbourne	Beijing
20	Milan	Frankfurt	**Dallas**	**Atlanta**
21	**Boston**	Berlin	Tokyo	Toronto
22	Taipei	Munich	Brussels	Osaka
23	Berlin	Hamburg	**Houston**	**Dallas**
24	Shanghai	Dusseldorf	**Philadelphia**	Sydney
25	**Atlanta**	Vienna	Berlin	Shenzhen
26	Vienna	London	Amsterdam	São Paulo
27	Munich	Edinburgh	Stockholm	Istanbul
28	**San Francisco**	Amsterdam	Munich	Moscow
29	**Miami**	Tokyo	Vienna	Barcelona
30	Brussels	Osaka	Osaka	**Houston**
Other US cities				
	Houston (34)			**Philadelphia (34)**
	Dallas (35)			**San Francisco (38)**
	Washington, DC (36)			**Washington, DC (41)**
				Boston (49)

* Aggregate of all indicators; US cities are in bold.

Source: MasterCard Global Centers of Commerce 2008.

Table 5.2 Top thirty cities for selected indicators and subindicators, 2008

Rank	*GCOC Index*	*Dealing with licenses*	*Registering property*	*Trading across borders*	*Political and legal frameworks**	*Quality of life*	*Basic services*
1	London	Copenhagen	Riyadh	Singapore	Stockholm	**Los Angeles**	Singapore
2	**New York**	Seoul	Stockholm	Copenhagen	Copenhagen	Sydney	Copenhagen
3	Tokyo	Stockholm	**New York**	Hong Kong	Singapore	**San Francisco**	Munich
4	Singapore	Singapore	**Chicago**	Stockholm	**Atlanta**	Melbourne	Frankfurt
5	**Chicago**	Frankfurt	**Philadelphia**	Frankfurt	**Boston**	London	Vancouver
6	Hong Kong	Berlin	**Los Angeles**	Berlin	**Washington, DC**	**New York**	Dusseldorf
7	Paris	Munich	**Boston**	Munich	**Chicago**	Paris	Tokyo
8	Frankfurt	Hamburg	**Atlanta**	Hamburg	**Houston**	Milan	Zürich
9	Seoul	Dusseldorf	**Miami**	Dusseldorf	**Los Angeles**	Rome	Stockholm
10	Amsterdam	Toronto	**San Francisco**	Tel Aviv	**Miami**	**Boston**	Vienna
11	Madrid	Montreal	**Houston**	Vienna	**New York**	Berlin	London
12	Sydney	Vancouver	**Dallas**	Amsterdam	**San Francisco**	**Washington, DC**	Osaka
13	Toronto	**New York**	**Washington, DC**	**New York**	Montreal	Vancouver	Montreal
14	Copenhagen	**Chicago**	Zürich	**Chicago**	Toronto	Tokyo	**Dallas**
15	Zürich	**Philadelphia**	Geneva	**Philadelphia**	Vancouver	**Chicago**	Paris
16	Stockholm	**Los Angeles**	Dubai	**Los Angeles**	Berlin	Vienna	Sydney
17	**Los Angeles**	**Boston**	Singapore	**Boston**	Frankfurt	**Dallas**	Toronto
18	**Philadelphia**	**Atlanta**	London	**Atlanta**	Munich	Dusseldorf	**Atlanta**

Table 5.2 Continued

Rank	*GCOC Index*	*Dealing with licenses*	*Registering property*	*Trading across borders*	*Political and legal frameworks**	*Quality of life*	*Basic services*
19	Osaka	**Miami**	Edinburgh	**Miami**	London	Johannesburg	Hamburg
20	Milan	**San Francisco**	Bangkok	**San Francisco**	Tokyo	Frankfurt	Amsterdam
21	**Boston**	**Houston**	Amsterdam	**Houston**	Zürich	Toronto	**Philadelphia**
22	Taipei	**Dallas**	Toronto	**Dallas**	Geneva	**Atlanta**	**Boston**
23	Berlin	**Washington, DC**	Montreal	**Washington, DC**	Vienna	**Miami**	Brussels
24	Shanghai	Mexico City	Vancouver	Seoul	Melbourne	Brussels	**Washington, DC**
25	**Atlanta**	Bangkok	Santiago	Dublin	Sydney	Amsterdam	Geneva
26	Vienna	Paris	Copenhagen	Tokyo	Amsterdam	**Philadelphia**	Melbourne
27	Munich	Dublin	Istanbul	Osaka	Barcelona	Osaka	**New York**
28	**San Francisco**	Zürich	Moscow	Toronto	Madrid	Munich	**Chicago**
29	**Miami**	Geneva	St. Petersburg	Montreal	Dublin	**Houston**	Berlin
30	Brussels	Brussels	Shanghai	Vancouver	Dubai	Barcelona	**Los Angeles**
Other US cities							
	Houston (34)						**Miami (31)**
	Dallas (35)						**Houston (32)**
	Washington, DC (36)						**San Francisco (33)**

Notes: US cities are in bold. *This indicator comprises more than the preceding set of subindicators.

Source: MasterCard Global Centers of Commerce 2008.

Table 5.2 Continued

Rank	*Banking services*	*Investor protection*	*Corporate tax burden*	*Contract enforcement*	*Ease of doing business**	*Financial services network: banking/ financial services companies*	*Starting a business*
1	London	Singapore	Singapore	Singapore	Singapore	London	Sydney
2	**New York**	Hong Kong	Hong Kong	Hong Kong	Hong Kong	**New York**	Melbourne
3	Singapore	Kuala Lumpur	Dubai	Vienna	London	Tokyo	Toronto
4	Hong Kong	**New York**	Riyadh	Sydney	Toronto	Hong Kong	Montreal
5	Paris	**Chicago**	Dublin	Melbourne	**New York**	Frankfurt	Vancouver
6	Frankfurt	**Philadelphia**	London	**New York**	Dublin	Singapore	Dublin
7	Amsterdam	**Los Angeles**	Edinburgh	**Chicago**	Edinburgh	Paris	Brussels
8	Madrid	**Boston**	Copenhagen	**Philadelphia**	Vancouver	Shanghai	Singapore
9	Copenhagen	**Atlanta**	Santiago	**Los Angeles**	Montreal	Milan	Paris
10	Zürich	**Miami**	Stockholm	**Boston**	**Chicago**	Madrid	Stockholm
11	Stockholm	**San Francisco**	Toronto	**Atlanta**	**San Francisco**	Amsterdam	**New York**
12	Berlin	**Houston**	Montreal	**Miami**	Sydney	São Paulo	**Chicago**
13	Munich	**Dallas**	Vancouver	**San Francisco**	**Los Angeles**	Seoul	**Philadelphia**
14	Vienna	**Washington, DC**	Amsterdam	**Houston**	**Boston**	Moscow	**Los Angeles**
15	Dublin	Tel Aviv	Johannesburg	**Dallas**	**Washington, DC**	Beijing	**Boston**
16	Brussels	Toronto	Zürich	**Washington, DC**	Copenhagen	Sydney	**Atlanta**
17	Hamburg	Montreal	Geneva	Seoul	**Atlanta**	Zürich	**Miami**
18	Barcelona	Vancouver	Lisbon	Paris	**Miami**	**Chicago**	**San Francisco**
19	Dusseldorf	Dublin	Beirut	Dublin	Melbourne	Mumbai	**Houston**

Table 5.2 Continued

Rank	*Banking services*	*Investor protection*	*Corporate tax burden*	*Contract enforcement*	*Ease of doing business**	*Financial services network: banking/ financial services companies*	*Starting a business*
20	Geneva	Johannesburg	Sydney	Budapest	**Dallas**	Kuala Lumpur	**Dallas**
21	Dubai	London	Melbourne	Tokyo	Tokyo	Mexico City	**Washington, DC**
22	Edinburgh	Edinburgh	**New York**	Osaka	Brussels	Jakarta	Hong Kong
23	Tokyo	Tokyo	**Chicago**	Brussels	**Houston**	Brussels	Copenhagen
24	Sydney	Osaka	**Philadelphia**	Frankfurt	**Philadelphia**	Bangkok	London
25	Toronto	Brussels	**Los Angeles**	Berlin	Berlin	Geneva	Edinburgh
26	**Los Angeles**	Copenhagen	**Boston**	Munich	Amsterdam	Istanbul	Zürich
27	Osaka	Bogota	**Atlanta**	Hamburg	Stockholm	Munich	Geneva
28	Milan	Lisbon	**Miami**	Dusseldorf	Munich	Warsaw	Tel Aviv
29	Montreal	Mumbai	**San Francisco**	Moscow	Vienna	Toronto	Lisbon
30	**Dallas**	Santiago	**Houston**	St. Petersburg	Osaka	**Los Angeles**	Amsterdam
Other US cities							
	Washington, DC (31) **Chicago (41)** **Boston (42)** **Atlanta (43)** **Miami (44)** **San Francisco (45)** **Philadelphia (52)** **Houston (54)**		**Dallas (31)** **Washington, DC (32)**			**Atlanta (36)** **San Francisco (38)** **Boston (41)** **Miami (42)** **Houston (49)** **Washington, DC (50)** **Dallas (54)** **Philadelphia (60)**	

*This indicator comprises more than the preceding set of subindicators.

Source: MasterCard Global Centers of Commerce 2008.

Table 5.2 Continued

Rank	*Employing workers*	*Closing a business*	*Air passenger traffic*	*Air cargo traffic*	*Conventions/ exhibitions/ meetings*	*Commercial real estate development*	*Business center**
1	**New York**	Singapore	London	Hong Kong	London	Hong Kong	Hong Kong
2	**Chicago**	Tokyo	**Chicago**	Tokyo	Paris	**New York**	London
3	**Philadelphia**	Osaka	**Atlanta**	Shanghai	Berlin	Singapore	Singapore
4	**Los Angeles**	Toronto	Paris	Seoul	Vienna	São Paulo	Shanghai
5	**Boston**	Montreal	Tokyo	Frankfurt	Singapore	Toronto	Dubai
6	**Atlanta**	Vancouver	**Dallas**	Paris	Hong Kong	Buenos Aires	Tokyo
7	**Miami**	Copenhagen	**New York**	**Miami**	Prague	Vancouver	Paris
8	**San Francisco**	Amsterdam	**Houston**	Singapore	**New York**	Dubai	**New York**
9	**Houston**	Brussels	**Los Angeles**	**Los Angeles**	Istanbul	Milan	Amsterdam
10	**Dallas**	Dublin	Frankfurt	London	Munich	Rio de Janeiro	Seoul
11	**Washington, DC**	London	**Washington, DC**	Dubai	Shanghai	Seoul	Frankfurt
12	Singapore	Edinburgh	Madrid	Amsterdam	Barcelona	Tel Aviv	**Los Angeles**
13	Copenhagen	Seoul	Shanghai	Taipei	Dubai	Tokyo	Bangkok
14	Sydney	Taipei	Amsterdam	**New York**	Seoul	**Miami**	**Chicago**
15	Melbourne	Sydney	Beijing	**Chicago**	Madrid	Osaka	**Miami**
16	Hong Kong	Melbourne	Moscow	Bangkok	Tokyo	Bangalore	Taipei
17	Toronto	Hong Kong	Milan	Beijing	Bangkok	Madrid	Madrid
18	Montreal	**New York**	Toronto	Osaka	Moscow	Caracas	Milan

Table 5.2 Continued

Rank	*Employing workers*	*Closing a business*	*Air passenger traffic*	*Air cargo traffic*	*Conventions/ exhibitions/ meetings*	*Commercial real estate development*	*Business center**
19	Vancouver	**Chicago**	**Philadelphia**	Brussels	São Paulo	Istanbul	Beijing
20	London	**Philadelphia**	Rome	**Dallas**	Beijing	Moscow	**Atlanta**
21	Edinburgh	**Los Angeles**	Seoul	**Atlanta**	Milan	Shenzhen	Toronto
22	Tokyo	**Boston**	Hong Kong	Kuala Lumpur	Budapest	Kuala Lumpur	Osaka
23	Osaka	**Atlanta**	Munich	Milan	**Chicago**	Santiago	**Dallas**
24	Dublin	**Miami**	São Paulo	Sydney	Edinburgh	St. Petersburg	Sydney
25	Zürich	**San Francisco**	**San Francisco**	Shenzhen	**San Francisco**	Montreal	Shenzhen
26	Geneva	**Houston**	**Miami**	Bogota	Brussels	Amsterdam	São Paulo
27	Kuala Lumpur	**Dallas**	Barcelona	**San Francisco**	Amsterdam	**Chicago**	Istanbul
28	Bangkok	**Washington, DC**	**Boston**	**Philadelphia**	St. Petersburg	London	Moscow
29	Riyadh	Stockholm	Bangkok	Mumbai	Sydney	**San Francisco**	Barcelona
30	Santiago	Madrid	Mexico City	São Paulo	Dublin	Shanghai	**Houston**
Other US cities				**Houston (34)** **Washington, DC (38)** **Boston (45)**	**Boston (45)** **Los Angeles (47)** **Atlanta (51)** **Miami (54)** **Washington DC (61)** **Dallas (66)** **Houston (67)** **Philadelphia (70)**	**Boston (39)** **Atlanta (44)** **Philadelphia (46)** **Washington DC (47)** **Los Angeles (57)** **Dallas (60)** **Houston (68)**	**Philadelphia (34)** **San Francisco (38)** **Washington DC (41)** **Boston (49)**

*This indicator comprises more than the preceding set of subindicators.

Source: MasterCard Global Centers of Commerce 2008.

consolidation of a European region and an Asian region. Thus the much discussed shift from a uni-to a multi-polar world can also be deselected in these findings. Asia, not surprisingly, now has four of the twelve leading echelon of global cities—Tokyo, Singapore, Hong Kong, and Seoul. And, more surprising, with the ascendance of Madrid into the top echelon—joining London, Paris, Frankfurt, and Amsterdam—Europe now has five. In contrast, the United States is left with only two (New York and Chicago), as Los Angeles has fallen from tenth to seventeenth place, and Boston, Philadelphia, Miami, and Atlanta are in a lower echelon. Notwithstanding the power of the US economy, the US's global cities are clearly part of a larger global network with major centers of gravity in Europe and Asia—a distributed formation. Toronto, Sydney, Zürich, and Copenhagen close out the top echelon of fifteen.

A second remarkable finding is that with the expansion of the global economy since the 1980s, the number of cities that are now capable of delivering global city functions has continued to increase. Thus, cities that fell out of the global network have now become reintegrated. One example is Buenos Aires, which saw firms and professionals leave as its severe economic downturn and bankruptcy erupted in 2000; the ensuing years saw the largest sovereign bankruptcy in modern history, and yet it has now become reincorporated. This also brings to the fore that global cities are built, developed, made in parts. Much investment and effort from local governments and firms, both national and foreign, goes into these processes. Since global firms and markets need a vast network of spaces of operation, it is economically efficient to reintegrate cities that have been rebuilt as platforms for the global economy.

Third, the data also signal that there is no perfect global city—no city ranks at the top in all the 100 data points aggregated into seven variables, on which the study is based. There is sharp variability. This also points to the fact that global firms and markets need many global cities, despite the fact that some are rather elementary—better many cities than one, single perfect city, as could possibly have been the case in earlier empires, where the capital was *the* city of the empire. Thus New York, one of the most important global cities, ranks rather low in a whole range of variables, which is not going to shock anybody. But more surprising are some of the very low rankings of various aspects of London, now the top global city in the world. A somewhat different pattern is evident in many of the mega-cities of the global south that have now emerged as global cities. These tend to rank high in global corporate aspects and very low in social issues. Thus, Mumbai and São Paulo are financial and economic powerhouses, and have high rankings in financial indicators, but they have very low rankings in basic services and general social conditions and livability. Yet they are critical to the global economy.

Fourth, a new type of city is emerging as a significant platform for global firms and markets. These are cities in small countries, where foreign firms locate not to invest in these countries but to use them as a platform. They have all the resources, the best legal and political frameworks, and basic services for all. They are veritable global platforms. Dubai provides this in a spectacular fashion in the Middle East, as does Singapore in Southeast Asia. But what stands out here is how the European cities have ascended in this domain. Thus, Copenhagen, with its strong legal and political framework, is emerging as a platform. For investors to go to Copenhagen is not necessarily about investing in that relatively small economy. Frankfurt, which in the 1980s quickly rose to become one of the top financial centers in the world, also emerged as such a platform. Locating in Frankfurt was not about investing in the city. Amsterdam, which ascended to tenth, is another such platform in what is a very small country.

Fifth, Zürich, which has long been, and remains the capital of international private banking, has lost ground to Amsterdam and Madrid as a financial center. Madrid, on the other hand, is reinserting itself in its old imperial geography, with heavy investments throughout Latin America. This has given Spanish firms and investors a strong base for significant financial gains, which in turn have enabled the acquisition of, among others, one of the largest UK banks and the British Airports Corporation. It indicates the extent to which private banking in today's global economy is not quite the power base it was, and that old imperial geographies can be reactivated.

Why is this expansion of the worldwide network of global cities happening? The general consensus is that most people now live in cities. But that is far too general a reason to explain the fact that our global economy, increasingly based on knowledge capital that can circulate electronically, has produced a "systemic" demand for a growing number of global cities over the last twenty years. Let us recall that, as recently as the 1970s, many of our most powerful cities had become poor— New York City, London, and Tokyo among them— and some had actually gone bankrupt.

The network of global cities has expanded as more and more firms go global and enter a growing range of national economies. The main business center(s) in each of these economies begins to evolve as a global city—it is the bridge between global firms and markets, and the specifics of national economies. This then also explains the multi-polar character of the network of global cities. In all of these cities we see the rebuilding of central areas—whether downtown and/or at the edges—as part of this new economic role. It amounts to rebuilding cities as platforms for a rapidly growing range of globalized activities and flows, from economic to cultural and political. This also explains why architecture, urban design, and urban planning have all become more important and visible in the last two decades.

Of interest to the question of megaregions in the case of the US is the fact that about half of the megaregions identified by Goldfeld (2007) in the United States contain at least one, but typically more, of these global commercial hubs. Beyond New York and Chicago, the overall GCOC Index positions the remaining four top US cities between the seventeenth and the thirtieth rank. Further, Houston and Washington, which in the earlier study were within the top twenty-three, are now well below that thirtieth rank. For Table 5.2, I selected only a few subindicators and indicators to bring to the fore: (1) the strengths of cities such as Atlanta and Philadelphia, usually not listed in the top four US cities; and (2) the weaknesses of the top cities, notably New York. The purpose is to illustrate the strengths and weaknesses of US cities in this global urban infrastructure for corporate capital, and the considerable variability of these strengths and weaknesses once we isolate specific criteria. It brings to the fore the ascendance of a distributed format, rather than simply a hierarchy as in older urban systems, and the fact that what matters is a network of cities, not just one (see also Annals of the American Academy of Political and Social Science 2009).

How these diverse, often disjunctural trends play out in the two other major regions, Asia and Europe, is a critical part of the overall picture I am seeking to specify. The experts of these regions and diverse cities are the ones most capable of addressing this.

Global intercity geographies: from hierarchical to transversal

One way to think about the global economy is in terms of the many highly specialized circuits that compose it. Different circuits contain different groups of countries and cities. Viewed this way, the global economy becomes concrete and specific, with a well-defined geography. Globally traded commodities—gold, butter, coffee, oil, sunflower seeds—are redistributed to a vast number of destinations, no matter how few the points of origin might be. The expansion of global trade has brought hundreds and hundreds of locations into these networks.[14]

These circuits are multidirectional and criss-cross the world, feeding intercity geographies with both expected and unexpected strategic nodes. For instance, New York is the leading global market to trade financial instruments on coffee, even though it does not grow a single bean. But a far less powerful financial center, Buenos Aires, is the leading global market to trade financial instruments on sunflower seeds. Cities located on global circuits, whether few or many, become part of distinct, and often highly specialized, intercity geographies. Thus if one were to track the global circuits of gold as a financial instrument, it is London, New York, Chicago, and Zürich that dominate. But if one were to track the direct trading in the metal, Johannesburg, Mumbai, Dubai, and Sydney all appear on the map.

In the last twenty years, there has been a sharp escalation in the capacity to control some of these economic circuits. Thus, while the commodities themselves may come from well over eighty countries and are sold in all countries of the world, about twenty financial exchanges control the global commodities futures trading. The map tightens when the seventy-three commodities traded are aggregated into three major groups. Five major global futures exchanges (NYME, LME, CBOT, TCOM, and IPE) are located in New York, London, Chicago, and Tokyo, and concentrate 76 percent of trading in these seventy-three commodities futures traded globally. Aggregated into three major groups, one single market clearly dominates in each. For agricultural commodities futures, the CBOT (Chicago) controls the most global trading; for energy, it is the NYME (New York); and for metals, it is the LME (London).

This tighter map of commodities futures trading illustrates the role of cities in today's globalizing and increasingly electronic economy. This is one of the points where global cities enter the picture. They are not the places where commodities are produced, but they are the places where commodity futures are invented, to facilitate the global trading of these commodities and help in managing some of the associated risks. This tighter map also makes concrete what is one of the main counter-intuitive trends evident in today's global economy—that the more globalized and non-material the activity (trading in financial instruments), the more concentrated the global map of those activities.

There are other such global maps, beyond commodities, commodities futures, and, in general, finance. Here, my focus on the global networks of global service firms will show us yet another version of this mix of globally distributed operations and high agglomeration economies. One way of tracking the global operations of firms is through their overseas affiliates. The top 100 global service firms have affiliates in 315 cities worldwide. For all multinational firms, the figure jumps to 1 million overseas affiliates.

The global operations of specialized service firms give us a sort of mirror image of the sharp concentration of the financial futures exchanges mentioned above. Their central operations are about managing the sales of their services to as many places as possible, whereas futures trading markets seek to concentrate as big a share of the trade in commodities as possible. The servicing operations of these firms become increasingly in demand in a growing number of cities as globalization expands. When countries open up to foreign firms and investors and allow their markets to become integrated into global markets, it is often the foreign service firms that take over the most specialized servicing; it has happened in cities as diverse as Buenos Aires and Beijing.

But the data point to a new pattern emerging along with this expanded globalization. It is the rise of cities such as Mumbai and Dubai as centers for servicing regional operations. More generally, the data show more transversal connectivities—that is, among mid-range cities emerging as service centers for other mid-range cities, rather than all international servicing originating in the leading global cities, as was the case in the 1980s and much of the 1990s. Returning to the issue of megaregions in the US, this growth of transversal mid-range connectivities points to an opening up of possibilities for mid-range cities in the US.

The global map produced by the operations of the top 100 service firms is dramatically different from the map produced by the financial trading of commodity futures, which is in turn different from that of the trading in the actual commodities. The extreme concentration apparent in finance would stand out even more if a map was drawn showing the trading of goods and the innumerable criss-crossing circuits connecting points of origin and destination. It does suggest that the specialized services are a sector that seeks out cities—as far as these firms are concerned, the more the better.

Implications for megaregions[15]

There is much to be said about megaregions. My concern in this chapter has been to identify possible types of intra-regional economic interactions that could be enabled by the megaregional scale. I focus on two key issues: first, whether such interactions might mean moving beyond economic planning styles that search to support and/or recruit only the most advanced economic sectors, such as "knowledge economies," and beginning to find value in the less developed areas of a region and in lower-income jobs; second, whether this co-presence of high- and low-value areas might in turn give megaregions particular advantages in today's global economy. If both of these are indeed the case, it is useful to ask whether and how novel types of megaregional coordination and governance could help in extracting and realizing these advantages. I have argued here that the central effort should be to understand the benefits of megaregions as going beyond familiar scale economies—that is, beyond the advantages of bigger and more of the same, whether mega-sized parking lots or longer train routes.

One component of this is whether a megaregion can seek to accommodate a larger range of the operations constituting a firm's value chain—from high agglomeration sites to dispersal sites. Practically speaking, this points to the possibility of bringing into or back to a megaregion some of the services and goods-producing jobs and operations currently offshored to take advantage of lower wages and fewer regulations. Can these be reinserted in the low-growth, low-cost areas of a megaregion? What type of planning would it take, and can it be done in ways that optimize the benefits for all involved—not only firms but also workers and localities? This would expand the project of optimizing growth beyond office parks and science parks, the preferred options today, and move across far more diverse economic sectors. It would use the lever of the megaregional scale to provide diverse spaces catering to different types of activities, ranging from those subject to high- and those subject to low-agglomeration economies. And, finally, the megaregional scale would help in optimizing the growth effect arising from the interactions of some of these diverse economies. This growth effect would be optimized by re-regionalizing some of the low-cost operations of firms today spread across the country and/or the world.

This way of thinking about the megaregional scale raises the importance of planning and coordination to secure optimal outcomes for all parties involved, including the challenge of securing the benefits firms pursue when they disperse their operations to low-wage areas. This would work for some types of economic sectors and types of firms, but not for all. Some activities that have been outsourced to other countries have not worked out, and have since been repatriated—they range from airline sales agents to particular types of design work in industries as diverse as garments and high-tech. But many of these outsourced activities are doing fine as far as the firms are concerned. Research and specific policies would be needed to establish the what, how, and where of the advantages for the pertinent firms of accessing low-wage workers within the US. This includes understanding how the location of these low-cost components in the megaregion where a given firm is headquartered could compensate for higher costs. It may require megaregional investment in developing low-cost areas for such jobs—a kind of rural enterprise zone.

There can possibly be a positive macro-level effect from repatriating some of these jobs, if a race to the bottom can be avoided and a certain level of consumption capacity is secured in the low-income areas of a region by ensuring reasonable wages or certain kinds of indirect subsidies. This brings a specific positive effect for a megaregion's less developed areas insofar as lower-wage households tend to spend a much larger share of their income on their place of residence—they lack the investment capital of the upper income strata, who can manage to allocate most of their income on overseas investments. Finally, this is also one element in the larger challenge of securing more equitable outcomes (for an analysis of options see, for example, Henderson 2005). It is important to ask about the distributive effects of the current configuration and of (potentially) optimized outcomes as described in this chapter. There is sufficient evidence on how extreme ill-distribution of the benefits of economic growth is undesirable in the long run.

These ways of specifying the meaning of a megaregion (or a region) take us beyond uses of the term as a sort of conceptual "packaging" to a more dynamic concept of the megaregion. Besides the advantages of urbanization, a megaregion is a sufficiently large in scale to optimize the benefits of diverse and interacting low- and high-income areas. What the megaregion offers in this context is a bigger range of types of locations than a city or a metro area—it offers locations tendering high-agglomeration economies all the way to locations where the advantage comes from dispersal. This would mean a direct growth effect between a megaregion's high- and low-agglomeration sites—the more the former grows, the more the latter will also grow. Then it becomes desirable for a megaregion to maximize the co-presence of these two types of locations. It also means freeing the lower-income area from its policy designation as a hopeless economic laggard.

In practical terms, there are, clearly, massive challenges for a megaregion to achieve this type of co-presence—maximizing the extent to which a megaregion can contain both the agglomeration and dispersal segments of a firm's chain of operations. For one, it is a counter-sensical, counter-intuitive proposition. It is not easy to see why a megaregion's highly dynamic economic spaces (the central areas of its global cities and silicon valleys), anchored by the headquarters of global and national firms, might actually be fed and strengthened in part by developing the "dispersal locations" of those same firms. Thinking of developing such "dispersal locations" as one way of making the most of negative externalities might make it more acceptable to the skeptics—you might as well go for activities that benefit from geographically dispersed arrangements once you hit excess congestion disadvantages. But one option at this point is, of course, to develop such items as golf courses and exurban oversized luxury housing. This is an argument that can be countered, since megaregions tend to contain much land that is not optimal for those uses but that could be optimal for developing "dispersal locations." Further, and critical to some of my substantive concerns for disadvantaged areas, these could benefit from such development—if a race to the bottom is to be avoided.

Notes

1 This chapter is derived from two larger projects that include policy discussions and extensive bibliographic materials (Goldfeld 2007; OECD 2007).
2 For one of the definitive examinations of the shortcomings of the data on sub-national scaling, see the report by the National Academy of Sciences 2003. See also OECD 2007.
3 Besides "regionalizing" various segments of a firm's chain of operations, one might also propose to regionalize more segments of various commodity chains. See, for example, Gereffi et al. 2005.
4 See, for instance, the forthcoming Special Issue of *Global Networks* (Derudder and Witlox 2010).
5 For one of the best data sets on the dispersal of the operations of firms in corporate services at the global scale, see GaWC n.d.
6 A parallel issue here, not fully addressed in this chapter, is the articulation of technical connectivity with social connectivity (Sassen 2008b: Chapter 7).
7 This is a type of agglomeration economy I found during my research on global cities, but it can also be applied to national or regional scales. The hypothesis is that the greater a firm's capacity for geographic dispersal, the higher the agglomeration economies it is subject to in its top-level headquarter functions (for a brief explanation of the nine hypotheses that specify the global city model, see Sassen 2001: New Preface). It is the specialized functions of globalized firms that evince the highest agglomeration economies. Their complexity, the uncertainty of the markets such firms are involved in, and the growing importance of speed in all these transactions are a mix of conditions that constitute a new logic for agglomeration; it is not the logic posited in older models, where weight and distance (cost of transport) are seen to shape agglomeration economies. I argue that the mix of firms, talents, and expertise in a broad range of specialized fields makes a certain type of dense environment function as a strategic knowledge economy, wherein the whole is more than the sum of (even its finest) parts.
8 This spatial lens is also to be distinguished from the more common angle of firms and markets (see, for example, Ernst 2005).
9 For a full development of this subject please see Sassen 2008b: Chapters 5 and 7; also Sassen 2001.
10 For a detailed explanation of this thwarting of technical logics by the economic, financial, or, for that matter, cultural and political logics of users, see Sassen 2008b: Chapter 7.
11 Today's multinationals have over 1 million affiliates worldwide. Affiliates are but one mode of global operation. For empirical details about some aspects see Taylor 2004; World Federation of Exchanges 2009 (annual updates).
12 For a detailed examination of the importance of the sub-national scale for a global market, see Harvey 2007.

13 The members of the research group are: Professor Fan Gang, Director, National Economic Research Institute, Beijing; Manu Bhaskaran, Partner/Head, Economic Research, Centennial Group, Singapore; Dr. Michael Goldberg, Professor Emeritus, Sauder School of Business, University of British Columbia; Professor William Lever, Emeritus Professor of Urban Studies, University of Glasgow; Professor Maurice D. Levi, Chair of Bank of Montreal Professor of International Finance, University of British Columbia; Dr. Anthony Pellegrini, Partner/Director of the Urban and Infrastructure Policy and Finance Practice, Centennial Group, Washington, DC; Professor Peter J. Taylor, Co-Director, Globalization and World Cities Research Group and Network, Loughborough University, UK; and myself.

14 This networked system also feeds unnecessary mobilities, because the intermediary economy of specialized services thrives on any kind of mobility. Thus, in the UK case, a study by the New Economics Foundation and the Open University of London found that in 2004, the UK exported 1,500 tons of fresh potatoes to Germany, and imported 1,500 tons of the same product from the same country; it sent 10,200 tons of milk and cream to France, and imported 9,900 tons of the same dairy goods from France; and so on over a vast spectrum of items.

15 This portion of the chapter is based on a paper prepared for the RPA and presented at Princeton University's Institute for Regional Policy, and subsequently published in the Institute's papers series (Goldfeld 2007).

References

Annals of the American Academy of Political and Social Science (2009) Special Issue on "The Shape of the New American City," *The Annals of the American Academy of Political and Social Science*, 626(1):1–241.

Derudder, B. and Witlox, F. (2010) "World Cities and Global Commodity Chains: An Introduction," *Global Networks*, 10(1): 1–11. Special issue on World Cities and Commodity Chains.

Ernst, D. (2005) "The New Mobility of Knowledge: Digital Information Systems and Global Flagship Networks," in R. Latham and S. Sassen (eds.) *Digital Formations: IT and New Architectures in the Global Realm*, Princeton, NJ: Princeton University Press, pp. 89–114.

GaWC (n.d.) Globalization and World Cities Study Group and Network. Online at www.lboro.ac.uk/gawc/ (accessed December 15, 2009).

Gereffi, G., Humphrey, J., and Sturgeon, T. (2005) "The Governance of Global Value Chains," *Review of International Political Economy (Special Issue: Aspects of Globalization)*, 12(1): 78–104.

Goldfeld, K.S. (ed.) (2007) *The Economic Geography of Megaregions.* Princeton: Policy Research Institute for the Region, Woodrow Wilson School of Public and International Afairs, Princeton University and the Regional Plan Association.

Harvey, R. (2007) "The Sub-National Constitution of Global Markets," in S. Sassen (ed.) *Deciphering the Global: Its Spaces, Scales and Subjects*, New York and London: Routledge, pp. 199–216.

Henderson, J. (2005) "Governing Growth and Inequality: the Continuing Relevance of Strategic Economic Planning," in R. Appelbaum and W. Robinson (eds.) *Towards a Critical Globalization Studies*, New York: Routledge, pp. 227–36.

MasterCard Global Centers of Commerce (2008) *Worldwide Centers of Commerce Index 2008*, MasterCard Worldwide. Online at www.mastercard.com/us/company/en/insights/pdfs/2008/MCWW_WCoC-Report_2008.pdf (accessed December 15, 2009).

National Academy of Sciences (2003) *Cities Transformed: Demographic Change and Its Implications in the Developing World*, Washington, DC: National Academies Press, Panel on Urban Population Dynamics.

OECD (2007) "What Policies for Globalising Cities? Rethinking the Urban Policy Agenda". Proceedings from the OECD International Conference, 29–30 March 2007, Madrid, Spain. Online at www.oecd.org/document/16/0,3343,en_21571361_37673954_40078672_1_1_1_1,00.html.

SAIS Review (2009) Special Issue on "Cities," *SAIS Review*, 29(1): 1–173.

Sassen, S. (2001) *The Global City*, 2nd edition, Princeton, NJ: Princeton University.

—— (2008a) "Re-assembling the Urban," *Urban Geography*, 29(2): 113–26.

—— (2008b) *Territory, Authority, Rights: From Medieval to Global Assemblages*, Princeton, NJ: Princeton University Press.

—— (2008c) "Mortgage Capital and its Particularities: A New Frontier for Global Finance," *Journal of International Affairs*, 62(1): 187–212.

Taylor, P.J. (2004) *World City Network: A Global Urban Analysis*, New York: Routledge.

World Federation of Exchanges (2009) *Annual Statistics*, Paris: World Federation of Exchanges.

6 America 2050

Towards a twenty-first-century national infrastructure investment plan for the United States

Robert D. Yaro

Introduction: Regional Plan Association and America 2050

This chapter describes America 2050, a program initiated by the Regional Plan Association (RPA) in 2005 to create a national growth strategy and infrastructure plan for the United States. America 2050 is now a joint venture between RPA and the Lincoln Institute of Land Policy. A major goal of this program is to promote the development of bottom-up infrastructure and growth strategies for America's metropolitan regions and an emerging network of ten "megaregions." Since its establishment in 1922, RPA has prepared three comprehensive long-range regional plans for the New York metropolitan region. With 23 million people and a \$1.2 trillion economy, this region is America's largest metropolitan area and its first "mega-city." RPA is an independent, nongovernmental organization, and although it has no official status in planning for the growth and development of the New York region, over the nearly ninety years since it was established, most of its key policy and investment recommendations have been implemented. In 1996, RPA completed its Third Regional Plan, and since then, many, if not most, of its key infrastructure and city-building recommendations have been adopted and implemented. The New York region continues to face a number of challenges, including economic restructuring, congestion, a shortage of affordable housing, and the need to conserve its natural areas and public water supplies in the face of expected climate change. RPA remains focused on addressing these challenges.

However, in the mid-2000s, RPA's leadership reached the conclusion that New York could not continue to succeed unless there was a broader development and infrastructure strategy for the rest of the Northeast Megaregion, and indeed the entire United States. For this reason, in 2005, RPA initiated the America 2050 program, with the goal of creating a national growth strategy and infrastructure plan for the entire US. The building blocks for this plan are America's metropolitan regions, which are home to 80 percent of the US population. Most metropolitan areas are situated in "megaregions"—networks of metropolitan areas with shared economic sectors and linked infrastructure and natural systems. An estimated 70 percent of the US population, and an even larger share of US gross domestic product (GDP), can be found in these places.

Why a national growth strategy and infrastructure plan is needed

America faces a host of challenges in the coming century, all of which will have a profound impact on the nation's future growth and development: global economic restructuring; rising fuel and household costs; climate change; rapid population growth; and deteriorating infrastructure. All these require strategies to maximize the nation's continued prosperity, opportunity, and quality of life.

In the face of these challenges, though, America is flying blind. No national strategy exists to build and manage the infrastructure systems needed to sustain inclusive economic growth and the country's competitive position in the global economy, and to secure a healthy environment for its children and grandchildren.

America's global competitors are now racing ahead to build the infrastructure to ensure their full participation in a twenty-first-century economy. While America spends about 2.4 percent of its GDP on infrastructure, China and India are spending 9 percent and 4.6 percent of their GDPs, respectively[1] (Congressional Budget Office 2007; Reuters 2007; Singer 2007). Every American lives with the consequences of decades of underinvestment, including: congested and deteriorating highways; unsafe bridges; inadequate transit and intercity rail systems; delayed flights; and bottlenecked seaports. Delayed shipments and the costs of congestion place every business in the country at a disadvantage compared with overseas competitors.

This was not always the case. The growth and development of the US was due in part to far-sighted investments that built canals, railroads, power generation projects, bridges, and roads, and protected the nation's environmental heritage, including its forests, wetlands, coastlines, parks, drinking water, and clean air. America has a history of national plans that shaped development: the Gallatin Plan in 1808; Theodore Roosevelt's conservation plans in 1908; and the work of the New Deal-era National Resources Planning Board, which led to the Interstate Highway System in 1956. It is time to do it again.

RPA's America 2050 project is dedicated to advancing a new vision for the future of America's infrastructure. Just as the Interstate Highway System provided a road map for the country's growth fifty years ago, so the US needs a similarly ambitious vision now, but one that responds to the challenge of increased foreign competition, while cutting greenhouse gases and reducing its over-reliance on imported oil. The US continues to have great potential to compete and lead in the twenty-first-century economy, because of its vast human, natural, and technological resources, but to do so effectively and efficiently, it must respond to four central challenges.

The challenges facing the United States

Global competitiveness

There has been an outpouring of studies in the past several years documenting the challenges facing the US as an economic, cultural, moral, and political world leader in the twenty-first century. For all their differences, three critical themes emerge. First, the US, as a mature industrial society and leader of the post-Second World War free world, is losing its competitive edge in the global economy. Second, the vast infrastructure systems created throughout the twentieth century are at the end of their useful life, require major reconstruction, or are overcrowded and congested, undercutting America's ability to compete in both its traditional fields of dominance and in emerging arenas of importance. Third, recognition of the problem is not the challenge; rather, it is the lack of imagination, creativity, and most of all, political will and leadership to rethink the fundamental principles and institutional design of the policy-making and governing processes of the nation. To do so will require harnessing the resources and capacities of both the public and private sectors to fashion twenty-first-century strategies for sustained and sustainable growth.

Infrastructure

Over the past two centuries, the US built systems of transportation, water supply and protection, and energy production and transmission that supported its growth, and was the envy of the world. Recently, America has not created the new capacity needed to underpin the next generation of economic development, let alone maintain what it built. An astonishing series of bridge collapses and levee failures have underscored America's dangerous infrastructure deficit, estimated by the American Society of Civil Engineers (ASCE) at $1.6 trillion (ASCE 2005). In the nation's transportation system alone, the recent National Surface Transportation Commission estimated that $225 billion a year is needed over fifty years to bring the transportation system to a good repair and make necessary improvements for economic growth. From transportation to water and energy systems, the US needs investment and innovation to develop a more sustainable framework to reduce its reliance on fossil fuels, create capacity for economic growth, and make better use of its natural resources (Kavinoky 2008).

Recent actions underscore President Obama's commitment to building advanced twenty-first-century infrastructure systems. Shortly after his inauguration Obama called for creating two elements of a long-range national infrastructure investment strategy: a national High Speed Rail (HSR) system and a "smart electric grid." The 2009 American Recovery and Renewal Act (ARRA) stimulus legislation included down payments for these systems. Eight billion dollars was committed to the ARRA for initial development of the HSR system, and $4.3 billion was committed to the initial phases of the smart grid.

The response from the states and from the public to these initiatives has been overwhelmingly positive. In response to a request for proposals from the Federal Railroad Administration (FRA) for these funds, FRA received forty-five applications from twenty-four states for a total of $50 billion in requests. In addition, the Chairman of the US House of Representatives, James Oberstar, has proposed that the next long-range national transportation authorization should include an additional $50 billion for HSR construction. However, this increased amount will not come close to meeting the needs. In 2008, voters in the state of California approved $10 billion towards that state's HSR lines, which will fund only a small portion of the estimated $45 billion needed to complete the HSR system just within their state. America 2050 has advanced the proposal for a "Trans American Network" of HSR links that would be 14,000 kilometers in extent. The ultimate cost of this system is estimated by America 2050 to approach $500 billion.

Fairness and opportunity

The global economic transformation is deepening economic and social stratification, as working-class and middle-class families are finding it increasingly difficult to cover rapidly rising transportation, housing, and fuel expenses. At the regional level, access to economic opportunity is often segmented by geography. In many places, the best jobs have moved out from central cities, and are accessible only by car, while poverty remains concentrated in inner cities and outer suburbs. At the national level, the transformation is leaving vast regions of the nation lagging in job growth and income levels. Predominantly manufacturing and resource-based economies, including the older industrial cities of the Northeast and Midwest and rural farming communities in the Great Plains and Deep South, have seen decades of declining population and income.

Climate change and energy security

In 2008, oil prices reached $120 per barrel. While prices have declined as a result of the global recession, they are expected to rise above $100 per barrel when the global economy recovers. Therefore, it is safe to say that the era of cheap gasoline has passed. Meanwhile, the evidence of human impact on global climate change is now unquestioned in serious scientific and political circles, and its likely effects will significantly change the lives of Americans. Climate change will impose huge costs on local communities across America, affecting agriculture, tourism, insurance rates, safety, and homeowners' investments. It will have a significant impact on hydrological cycles, producing more frequent droughts and flooding, as well as other consequences. At the national and global scales, climate change will cause a disproportionate impact on developing countries, including vulnerable coastal regions in the Global South, contributing to competition for resources and instability in existing volatile regions. The Intergovernmental Panel on Climate Change (IPCC) estimates that the US must reduce its greenhouse gas emissions by 60 to 80 percent of its current levels by 2050 to avert the worst impact of climate change, with every year of delay driving up the cost of making change (IPCC 2007).

In face of these challenges, the American population is anticipated to grow rapidly. According to the US Census Bureau, from 2000 to 2050 America will grow by about 140 million additional people—the same number that was added from 1950 to 2000, during which time America's GDP increased almost six-fold. By making new investments in the physical fabric of the nation, America can create the capacity necessary for growth, in the same way that the Interstate Highway System fueled two generations of development in the latter half of the twentieth century, and a system of canals, roads, and railroads supported growth in the nation's first 150 years (Mahone 2007). The challenge for the present generation of Americans is to make these investments in the nation's physical infrastructure, including transportation and energy infrastructure, the natural environment and water resources, and the next generation of commercial, residential, and government buildings. With equally bold strategies for investing in people skills, these investments will provide the foundation for a new generation of more fair and sustainable development.

A history of national planning

Throughout the nation's history, Americans have met challenges even more difficult than those facing them today. Two centuries ago, the United States faced the test of building a continental-scaled economy and mobility systems, while simultaneously creating strong, physical links between the nation's diverse regions (Howes 2007). At the turn of the last century, the nation faced the challenge of building great cities to accommodate rapid immigration and industrialization, and the infrastructure systems needed to service what was becoming the world's largest manufacturing economy. Concurrently, they needed to take bold action to preserve the nation's dwindling natural resources and provide clean air and water to an increasingly urbanized society.

To address these needs, in every era since the founding of the nation, Americans have developed and implemented national infrastructure investment plans that have had a profound impact on America's growth and development. Historian Robert Fishman has documented three important national plans that profoundly influenced the shape of national development, often via indirect means. A brief history is in order to set the context for the ambitious proposals that are needed for the century before us (Fishman 2007).[2]

1808

At the request of President Thomas Jefferson, Treasury Secretary Albert Gallatin created a national plan of ports, roads, and inland waterways to encourage settlement of the nation and facilitate trade among independent farmers scattered across the land. This plan was inspired by the visions of George Washington and Thomas Jefferson of an egalitarian society—the "homestead vision" that was first made possible by the Land Ordinance of 1785. While the plan's implementation was slowed by growing north–south divisions between free and slave states, several states moved aggressively to implement Gallatin's vision. New York State, for example, built the Erie Canal, changing the geography of commerce in the nation. Other key elements of the Gallatin plan were realized under Abraham Lincoln's leadership in 1862, when Congress enacted the Homestead Act, which granted 160 free acres to any family willing to farm the land, and the Pacific Railway Act, which launched the first transcontinental railroad. Through this alliance with private interests —the railways and the homesteaders—a national transportation system was realized at a rapid pace and at an enormous scale, and it resulted in dramatic (but uneven) generation of wealth. The same year, inspired by the land grant principle underpinning the Gallatin Plan, Congress adopted the Morrill Land Grant Colleges Act, creating the nation-wide system of state universities, which dramatically expanded and democratized the nation's higher education system.

1908

Theodore Roosevelt convened a Conference of Governors at the White House to launch a series of conservation plans drafted by the Inland Waterways Commission that would target underperforming regions of the South and West with irrigation, river restoration, and dam projects. The plans responded to a generation of railroad development that had consolidated power and wealth in the hands of private interests concentrated in the Northeastern and Midwestern cities. The conservation programs sought to restore ravished farmland and clogged river corridors and create cheap power with hydroelectric dams, resulting in successful projects such as the Roosevelt Dam near Phoenix (1911), the Colorado Compact (1922), and the Hoover Dam (1931). Twenty-five years later, the ideas of the Inland Waterways Commission were taken up under the New Deal by the Tennessee Valley Authority, Bonneville Power Administration, and other conservation and economic development programs under President Franklin Roosevelt.

1930s and 1940s

The National Resources Board (later renamed the National Resources Planning Board, or NRPB) was established by President Franklin Roosevelt in 1933 to guide infrastructure investments promoted by the economic stimulus programs of his Depression-era New Deal. Led by a distinguished board chaired by Roosevelt's uncle, Frederick Delano, the Board's assignment was to take full account of "the distribution and trends of population, land uses, industry, housing, and natural resources" across regions, river basins, and entire ecosystems, to assure that federal expenditures would deliver "comprehensive and coordinated" consequences. The Board performed crucial services throughout the 1930s in coordinating public works and forming policy for land use, forests, parks, hydrology, and government-sponsored defense-plant location. Perhaps the Board's most important contribution to the nation's infrastructure system was its plans for a national toll road and free road system, which became the basis for the postwar development of the Interstate Highway System advanced by President Dwight Eisenhower.

The impact of the Interstate Highway System was immense. Coinciding with the return of Second World War veterans, and federal programs that encouraged home ownership, it spawned patterns of suburban development that define the nation's character today. As was the case in the railroad era, the convergence of federal incentives and private investment resulted in the completion of an astonishing transportation network, the unification of the country, and uneven costs and benefits, in this case between urban and suburban areas.

Rebuilding and renewing America in the twenty-first century

The transformative infrastructure investments of America's first two centuries shared a number of key characteristics. They inspired public investments that enabled private citizens to better access economic opportunity. They were built in cooperation with the private sector. They were national in scale. They relied on state and local actions to implement their national visions. They responded to the most urgent challenges of their time. Jefferson and Washington's Homestead vision of using land and infrastructure enabled Americans to pursue their economic interests, and at the same time build a nation. The visions of [the] Roosevelt(s) and Eisenhower for nationwide programs of conservation, infrastructure investment, and personal mobility opened up access to the American Dream, and built an industrial powerhouse.

Infrastructure investment as the framework for equitable and sustainable economic growth

In the past, national plans and the infrastructure investments they inspired resulted in dramatic economic growth. In many places, however, this growth was accompanied by unanticipated consequences, including damage to the environment and often unfair, inequitable distribution of economic benefits. In twenty-first-century American society, however, these tradeoffs should not be accepted as the inevitable consequence of economic growth. As in the past, the federal government should provide leadership in setting goals for national-scale infrastructure investments that will shape the quality and direction of the nation's growth, while simultaneously focusing on the steps needed to ensure that these investments promote more sustainable and more equitable patterns of development.

The concept of the "triple bottom line," introduced in the 1990s as an accounting method to capture the economic, ecological, and social impact of investments, captures the same values that should be considered in weighing major public infrastructure investments and policies. Instead of after-the-fact accounting, however, the federal government should set out performance standards that measure progress toward meeting economic, environmental, and social equity goals and can be used in the project decision-making process.

Looking at one of the nation's largest historic infrastructure investments, the Interstate Highway System, one could imagine an alternative form that would have resulted had each individual segment of the broader vision been subject to performance standards related to environmental sustainability and fairness. While the system as a whole would have largely the same scope and positive impact on national economic growth, individual components may very well have looked dramatically different than they do today. New York's Cross-Bronx Expressway, as an example, is clearly integral to the highway network of the eastern seaboard, yet it has had a pronounced negative impact on the health, economic vitality, and environmental quality of the local community through which it passes. A stronger consideration of avoiding the adverse environmental and social impact would have identified strategies that included a slightly different course, decking, local parkland, housing replacement, and other elements that would have enabled the same level of regional and national economic development while mitigating many of the local negative externalities.

The role of the federal government in a national infrastructure plan

While many land use decisions and infrastructure investments in the US are made at the state and local levels, historically, federal policy and resources have had an unseen hand in shaping the development of the national landscape. For example, the combination of the Interstate Highway System and the lending practices of the Federal Home Loan Mortgage agency in the post-Second World War era transformed America into a more suburban nation. Today, federal policies and funding programs continue to shape America's growth, and not always in ways that help America respond to the key challenges of the twenty-first century (Appleton 2007).

With the urgent need to reduce America's reliance on foreign oil and mitigate and adapt to climate change, national policies should be aligned to support the type of development that will make the most of the nation's existing infrastructure systems and built areas, and provide increased mobility options. The national role in infrastructure investment, development, and environmental protection should be one that sets the direction of goals and objectives expressed as outcomes—e.g. livability, mobility, accessibility—that meet America's "triple bottom line" of economic prosperity, fairness, and sustainability. These outcomes, expressed as standards and performance measures, would provide a framework for challenging states and local levels of governments to meet these goals with financial incentives and technical support.

The national infrastructure investment plan could play a constructive role in national development in the following ways:

1. It could establish a vision and a set of priorities for the nation's infrastructure needs by determining goals to align departmental policies, ensure best value from government spending, reduce duplication, and demonstrate how investments can reinforce one another. This framework would avoid the enormous time delays and costs in delivering major projects that arise due to the lack of agreement upon national priorities, and would set the context within which decisions are taken at the state and local levels (and by the private sector) with confidence and evidence.
2. It could set standards for efficiency and safety (e.g. CAFE [corporate average fuel economy] standards for automobiles and light trucks, efficiency standards for appliances). In safety, the federal government plays an important role in many areas, including the Federal Aviation Administration, the Food and Drug Administration, and the Occupational Safety and Health Administration. The federal government should set a high bar for minimum standards of safety, efficiency, and health, and yet individual states should not be discouraged from aiming higher.
3. It could promote federal objectives with "conditionality." The promise of federal funding and financial tools for states and regions that take action to meet federal objectives can act as a powerful motivating force. Equally compelling is withholding federal funds or financing where federal objectives are not being met. For example, Congress used both the carrot and stick approach in the 1991 highway bill, to promote state laws mandating seat belts and motorcycle helmets.
4. It could convene multi-state partnerships. To address issues of a multistate nature, the federal government can provide the incentive and the legal basis for states to finance multi-state transportation projects or regulate and protect multi-state environmental resources. Entities such as the Port Authority of New York and New Jersey, the Alameda Corridor Authority, the Tennessee Valley Authority, and the Bay Delta Commission were created through federal actions.
5. It could measure performance and collect data. The federal government should collect data to measure and rate the performance of states and regions toward meeting the outcomes of federal programs for which they receive funding, as has previously been done in education, with the No Child Left Behind Act. Rigorous evaluation of performance should be extended to other federal programs, particularly transportation, to increase the transparency of how federal funding is used, and to provide the tools to make better decisions about the effectiveness of federal programs.

Interactions with states, regions, and local levels of government

Individual states and local governments will continue to play the important role of planning, developing, and maintaining much of the nation's infrastructure investments within the context of a national vision, clear federal priorities, and performance standards. Across the country, states have been leading the way in adopting climate change policies that are more ambitious than those of the federal government. For example, California's AB 32 legislation, which requires carbon dioxide reductions to 1990 levels by 2020, has already begun to foster innovation and investments in new technologies that will help the rest of the nation respond to the climate change challenge. To date, twenty-nine US states have also adopted Renewable Portfolio Standards (twenty-seven of them mandatory), which set varying targets for the states' use of renewable energy.

Metropolitan regions will also play an important role in transportation policy and in coordinating transportation and land use investments to promote greater energy efficiency, sustainability, and quality of life. Measured by the geographic extent of integrated labor markets, metropolitan regions function as the basic units of the nation's economy. In transportation policy, specifically, the case for strengthening the decision to give power and direct funding to regional authorities, such as Metropolitan Planning Organizations and regional transit authorities, is compelling. According to the Brookings Institution, America's top 100 metros alone account for over three-quarters of the nation's air cargo tonnage, seaport tonnage, and rail passengers, yet they struggle to receive adequate funding and tools from their respective state governments, despite contributing disproportionately to state and federal gas tax revenues.

Cities and local governments have important roles to play in concentrating jobs, housing, and activities in central places, where transportation options are plentiful. After enduring a long era of decline from the 1960s through the early 1990s, many of the nation's urban places have undergone radical transformations to become more stable, prosperous places, while others are just beginning to turn around. The nationwide decline in crime and sustained national economic growth has played a role in this transformation, as have demographic and market forces that are leading more young people, immigrants, and "empty nesters" to seek homes in urban places.

Megaregions: a framework for meeting cross-border challenges

In addition to states, regions, and local levels of government, a new urban scale has emerged that presents a framework for responding to large-scale challenges.

Megaregions are networks of metropolitan areas, connected by travel patterns, economic links, shared natural resources, and social and historical commonalities. America 2050 has identified ten emerging megaregions in the US, which are presently home to 70 percent of the US population, and where the majority of the nation's growth before 2050 will take place. The connections between the component metropolitan areas of megaregions are growing stronger as technological and global economic factors enable the movement of goods, people, and information at lower costs, more frequency, and at greater speeds (Sassen 2007). These megaregions are America's gateways to the global economy, where its global ports, airports, communication, financial, and marketing centers are located.

While megaregions present a new scale to consider and plan for, it is unnecessary to form megaregion "governments" to address emerging megaregional issues. Rather, adjacent states, cities, or metropolitan areas should recognize the issues taking place at this scale and form partnerships to address them. For this to occur, the federal government will need to undertake the leadership role described above.

Increasing interactions between the metropolitan areas that make up megaregions also present the opportunity to expand regional agglomeration economies to a larger scale. With better transportation connections, firms can take advantage of lower-cost labor and start-up costs in rural or older industrial communities that are distant from the economic core but still within the megaregion, as an alternative to international outsourcing. A megaregion can potentially take advantage of a wide range of economic functions (from headquarter locations to low-skilled manufacturing) within a single geographic "megazone" (Sassen 2007).

Supporting collective megaregion-scale actions could achieve the following:

1. development of intercity and high-speed rail corridors that move people and goods, which are linked to America's global facilities and other multi-state or interregional transportation networks, which require coordinated planning, investment, and management across local, state, and international borders;
2. protection, restoration, and management of large environmental systems and resources such as watersheds, farmland, forests, and coastal areas;
3. development of economic revitalization strategies for underperforming regions such as the Midwest and the Gulf Coast.

Three components of a national infrastructure investment plan—water, energy, and transportation

A national infrastructure investment plan should focus on a full range of infrastructure systems. While the principal focus of the plan should be on water, energy, and transportation systems, the plan should also address broadband communications and other systems.

Water

Despite, and perhaps because of, their many successes, the water resource policies of the last thirty years must be recast for the new century. Population growth, changing settlement patterns, aging infrastructure, unmet environmental goals, climate change, and global markets for agricultural commodities are posing distinct new challenges. America's water infrastructure was once an unrivaled advantage over other nations. If it is to remain so, water policy must change with the times, in ways that decisively address the aforementioned challenges and that make clean water a central element of a sustainable future.

Water resource management now crosses many political, economic, institutional, and disciplinary boundaries (O'Keefe et al. 2007). Creating guiding principles that work across them to turn these overlaps into a synergistic asset must be must be a top priority for a national infrastructure plan. Water resource managers increasingly understand that the heavily engineered, capital-intensive, facility construction solutions that characterized twentieth-century approaches to water supply, wastewater treatment, storm water management, and flood control problems are no longer sufficient. Now, an increasing complex set of water problems requires a policy framework that encourages innovation and efficiency, understands resource optimization, promotes creative uses of market tools, eliminates perverse incentives and subsidies, aggressively promotes watershed management, and links water resource management to land use decisions (*Milwaukee Journal Sentinel* 2008).

The landmark 1972 Clean Water Act and subsequent federal and state policies allowed the United States to make enormous strides in reducing pollution from municipal and industrial point sources. According to the Environmental Protection Agency (EPA), pollutant discharges are half of what they were in 1970, despite waste loads that grew by one-third. Cleaner water has benefited public health, created new recreational opportunities, improved fisheries and wildlife habitat, and subsidized urban growth and development. These achievements are directly tied to the level of investment in water resource infrastructure and water resource management since 1970. The EPA estimates that public spending on water systems has doubled since then and spending on wastewater systems has tripled (United States Environmental Protection Agency 2002; Howes 2007).

Despite these accomplishments and expenditures, the Clean Water Act goals of fishable and swimmable waters have not been universally met. The most serious remaining problems occur in the bays and estuaries where older combined sewers, diffuse non-point urban storm water, and growing agricultural runoff have remained largely uncontrolled by traditional capital investment approaches. It is now generally recognized that restoration of these water bodies, such as Puget Sound, San Francisco Bay-Delta, the Everglades, Chesapeake Bay, and the Gulf of Mexico, require complex multi-stakeholder strategies and investments.

Population growth and migration are also changing the priorities of water resource management in ways that national water infrastructure policy needs to recognize and address. Emerging megaregions and growing metropolitan areas will significantly increase demands for water in these urbanized areas. Physical scarcity of water is a growing concern in Southern California, the

Arizona Sunbelt, and the Texas Triangle. The Piedmont and Front Range region have adequate water but experience economic scarcity caused by lack of institutional, financial, and human capital (Mahone 2007). Even in seemingly water-rich locations such as the Northeast, ex-urban sprawl into water supply watersheds is requiring new infrastructure upstream, while degrading quality for downstream urban areas.

Climate change is expected to strongly alter hydrologic cycles and the dynamics of this debate. The reliability of water supplies will be reduced, especially in areas dependent on seasonal precipitation cycles. Warmer temperatures and changing precipitation patterns will reduce annual snow packs and increase evapotranspiration, lowering natural storage capacity. Weather extremes will occur with greater frequency, stressing not only humans, but also wildlife and natural systems. Sea-level rise and higher intensity storms increase risks on the coasts and inland. Flooding will incur even greater costs in addition to the billions of dollars of damage the nation experiences each year, compounded by land use decisions that allow construction in flood-prone areas and a reliance on levees and other engineering structures built in opposition to, rather than as a complement of, natural hydrological dynamics. Federal and state flood insurance and disaster relief policies have allowed developers to avoid the financial consequences of such unwise decisions (Otto 2008).

The nation's water infrastructure is aging, too. Many urban areas still rely on distribution systems installed in the nineteenth and early twentieth centuries. The massive infrastructure investments of the 1970s are beginning to near the end of their useful life and will soon require renovation or replacement.

Addressing these needs will be expensive. While federal, state, and local sources have invested about $18 to $20 billion a year towards these capital needs since 1970, the EPA has projected a gap of $224 billion, or about $11 billion a year, to meet replacement costs and unmet goals over the next twenty years. The relative role of different levels of government, utilities, and developers in closing this gap and the affordability for ratepayers and taxpayers are key issues.

Addressing all these challenges will require a more integrated set of solutions. The EPA has outlined a "four pillars" approach that incorporates principles of better management, full cost pricing, efficiency, and watershed approaches. Two recent reports by the National Academy of Public Administration have also proposed that the EPA and the Army Corps of Engineers develop outcome-orientated policies and integrated strategies that connect land use and investment actions.

The rationale for promoting non-structural solutions to water resource problems is clear. Pollution prevention, water conservation, appropriate pricing, resource optimization, and improving management efficiency can provide the same benefits at a far lower cost than capital intensive facilities. These approaches have also been proven to have the added advantage of providing important secondary benefits, including saving energy, protecting and restoring ecosystem services, lowering liability associated with flooding, and creating local jobs by relying on labor management-intensive activities.

To be effective, a national infrastructure investment plan must outline how traditional federal mandates and ongoing capital investments in water management will be linked to non-structural alternatives (Appleton 2007). It should provide incentives for smarter systematic approaches that link upstream and downstream investments, and match relative effectiveness to relative cost. Perhaps most critically, it should break down traditional sector responsibilities and deliver the coordination needed for water managers to engage land use decision-makers.

The national water policy choices that will be made over the next several years will determine whether America's water resource managers, in the face of growing challenges and complexity, can produce safe drinking water for over 300 million Americans, dispose of their sewage safely, provide industry and agriculture with the water it needs, and do all of this and much more in a way that is both environmentally sustainable and economically affordable. The task will be to use the successes of the past and the many promising initiatives of the present to create the integrated, multi-dimensional, goal-oriented policies that the future holds the responsibility for (Allbee 2005a).

Energy—technologies, climate change, and governance

America's response to the dual challenge of meeting its growing energy needs and responding to the threat of global climate change will define its ability to compete globally in the twenty-first century. Generating and delivering electricity to meet the demand of America's growing "hi-tech" society is a critical component of its national infrastructure (Friedman 2008). Without the capabilities to provide an adequate supply of energy to meet demand, its economy will stagnate or contract as businesses move to where their energy-intensive needs can be met. And yet, with the nation's energy demand growing at a faster rate than its population, attention to managing demand by promoting greater efficiency in new and existing buildings, industry, and transportation must be at the core of national policy. To meet this challenge, America must holistically examine its existing electrical power generation, transmission, and distribution infrastructure, much of which still relies on antique nineteenth-century technology. America must identify areas for investment, assign priorities, and clearly define the roles and responsibilities of both the private and public sectors.

As the world's largest economy, the US also has the responsibility to provide global leadership in reducing greenhouse gases through the adoption of renewable sources of generation. The nation currently emits approximately twenty-four tons of greenhouse gases per capita, compared to China's three tons per capita. With China looking to emulate America's post-Second World War growth patterns and automobile culture as it embarks on a period of unprecedented economic and population growth, the need for America to show leadership in climate change responsibility is clear. China, like the US, relies heavily on coal-fired power plants for electricity generation, a major source of greenhouse gas emissions. Domestically, electricity production is responsible for 40 percent of US carbon emissions—a direct result of America's heavy reliance on coal as a fuel source. The share of US electricity generation from coal-fired power plants is projected to grow from 50 to 57 percent by 2030 (United States Department of Energy Office of Electricity Delivery and Energy Reliability 2008).

Over the next decade, energy efficiency offers the most immediate large-scale payback from investments. Retrofitting commercial, industrial, and government buildings and home weatherization will stop waste and cut demand. These programs require a clear commitment by the federal government to act on its own properties and practices and to provide tools and incentives for homeowners, businesses, and state and local governments.

Renewable energy, such as solar and wind energy, is seen as the wave of the future, but without sustained government investment, these technologies will be unable to compete with fossil fuels. Congress included $20.8 billion in the 2009 ARRA stimulus legislation to promote development of alternative energy systems, including solar and wind energy. A sum of $4.6 billion was also included in the ARRA to promote "smart" electric grids and to help bring wind and solar power sites in isolated rural regions to metropolitan markets (United States Congress 2007). Transmission will also have to be altered to be able to handle the distributed generation nature of small-scale solar energy.

Finally, there are a number of technologies that, if implemented, will have the potential to mitigate some of the environmental impact of the legacy power plants. Advancements in "clean coal" and carbon capture and sequestration (CSS) technologies could substantially reduce the damaging emissions from these plants, and create new marketable technologies to China and other developing counties. The ARRA included $3.4 billion to promote carbon sequestration demonstration projects and technology development. Investments in renewable energy sources, as well as a greater role for nuclear and natural gas power generation, could also be a part of a comprehensive package of improvements contributing to the reduction of America's national carbon footprint. In the near future, America must build the infrastructure necessary to secure an adequate supply of natural gas as a bridge to future technologies. Recent discoveries of extensive quantities of natural gas in the so-called Marcellus Shale belt, stretching from Tennessee to New York, promises to significantly increase US natural gas reserves.

The 2007 energy bill began to address these concerns by advancing research on the "Smart Grid" (Zheng 2007). This next-generation network, funded in part by $4.3 billion in ARRA stimulus funds, will integrate broadband telecommunications technologies and sensing devices that would be installed in homes and appliances, permitting bi-directional communication between the customer and energy supplier. These technologies would allow for real-time monitoring and pricing, and provide for automatic load adjustment on the supplier and consumer ends, which would increase throughput, conserve energy, and prevent blackouts (United States Department of Energy Office of Electric Transmission and Distribution 2004).

The "Smart Grid" will significantly improve the transmission system's ability to respond to and recover from an unforeseen crisis through its ability to "self heal," preventing a reoccurrence of large-scale outages such as the one experienced in the Northeast in the summer of 2003. Perhaps the greatest advancement of this new grid is that it will be designed to both distribute and store energy. Everyone will be able to draw or purchase power from the grid and feed or sell any surplus back to it. This surplus power can then be allocated to meet other demands. This model improves the economics of cogeneration—the simultaneous production of heat and electricity—and promotes the use of renewable energy. This innovative program was funded at $100 million each year, from 2008 to 2012 (Weinshel 2008).

Private sector corporations play a dominant role in the US energy sector, as they provide 75 percent of net power generation and own almost 80 percent of transmission lines. The remaining generation and transmission is provided by government power marketing authorities, such as the Tennessee Valley Authority (TVA) and other cooperatives. Because the private sector is positioned as the principal provider and owner of America's energy infrastructure, market forces tend to dictate whether infrastructure investments are promoted. To get around this issue, the government could further incentivize industry investments in renewable energy sources and in improvements such as the "Smart Grid."

The federal government has initiated the process of creating a national framework that ensures the stability of America's future energy supply, the adequate generation of electrical power to meet demand, a robust infrastructure to deliver electricity, and an ability to lead in fighting global climate change. By actively engaging both the public and private sectors, the US can systematically and comprehensively plan for the long-term health and efficiency of these critical systems.

Transportation—the role of a national plan

Meeting the needs of America's deteriorating transportation infrastructure and providing capacity for future, sustainable growth must be at the core of any national infrastructure plan (Allbee 2005b). The current surface transportation bill, SAFETEA-LU (Safe Accountable, Flexible, Efficient Transportation Equity Act: A Legacy for Users), expired in September 2009, and industry associations, policymakers, and advocates are now focusing on shaping the next bill. Widespread among recent reports and statements by groups such as the National Surface Transportation Policy and Revenue Study Commission, American Association of State Highway and Transportation Officials (AASHTO), the American Road and Transportation Builders Association (ARTBA), the American Public Transit Association (APTA), the US Chamber of Commerce, and the Brookings Institution are three themes: (1) the recognition that the current policy framework is not meeting the nation's transportation needs; (2) the incredible funding gap to repair and expand the system; and (3) the call for a clear "vision" to guide transportation investment and the federal role in the twenty-first century (AASHTO 2007; ARTBA 2007; National Surface Transportation Policy and Revenue Study Commission 2007).

In addition to these broad points of agreement, calls for real reform of transportation policy are growing. The need for a "new beginning" was clearly stated in Transportation for Tomorrow, the final report of the National Surface Transportation Policy and Revenue Study Commission, which recommended not "reauthorizing" the current transportation bill in favor of crafting an entirely new transportation policy (Downey 2007; Puentes 2008). Its characterization of the nation's transportation program since completing the Interstate Highways system as "pursuing no discernable national interests other than the political imperative of 'donor states' rights and congressional earmarking," makes clear the failure of today's policies to achieve national goals. In fact, the Commission charged that today's transportation policies work against national goals, by perpetuating oil dependency, raising greenhouse gas emissions, and undermining economic competitiveness, national security, public health, and safety.

Despite the clarity of the Commission's criticism, engaging Americans and politicians in a debate about the minutiae of transportation policy reform is a challenge. The media and political frenzy in 2008, inspired by gasoline that cost four dollars per gallon, did little to promote a substantive debate about how the highway trust fund is spent, for example. Instead, it garnered short-term proposals for a gas tax "holiday," which clearly would do nothing to address the more fundamental problem—that most Americans have no other choice in getting around but to drive. The trouble is that in response to rising congestion and high gas prices, national transportation policy is engineered to promote a single solution above others—build more roads—which does little to address either problem.

Clearly, more investment is needed. But, drawing from the same example, the transportation investments of tomorrow should proceed on a level playing field, where communities are offered the same tools and resources to invest in public transportation, transit-oriented development, biking, and walking, as they are offered to build roads. More importantly, this new approach should be integrated in federal goals and performance objectives that add up to a vision for national transportation policy. With broad strokes and illustrative examples, a national transportation vision for investment has the power to capture the general public's imagination in the way that detailed reform proposals may not (Puentes 2008).

The precedent of the federal government's leadership in planning and building the nation's Interstate Highways System in the twentieth century is frequently cited as a model for planning the bold, national-purpose transportation investments of the twenty-first century. The challenges of the twenty-first century will call for a new and more active role from the federal government, in establishing principles, priorities, and plans to guide transportation investment in the dawning century.

First and foremost among these is the growing role of foreign trade in the nation's economy and its growing impact on international ports, key trade corridors, and intermodal chokepoints. Currently, states and local governments lack the resources, multimodal scope, and ability to coordinate across jurisdictions to make effective and prioritized investments in key global gateways and critical trade corridors. Moreover, with the growing imperative to reduce greenhouse gas emissions, polluting port facilities and their transportation connections must begin to make the transition to low-carbon practices, which will require federal performance standards and investments.

The second trend is the emergence of megaregions—linked networks of metropolitan regions that are experiencing increased international trade, interregional goods movement, and intercity passenger travel. Without an intermediary level of government between states and the federal level, megaregions are relying on ad hoc collaborations to strategize and prioritize projects in key multi-state corridors.[3] A stronger federal role in convening megaregion partnerships to plan goods movement corridors and high-speed rail networks, for example, could begin to fill this gap and is a logical focus of a national infrastructure plan.

The third trend is the converging concern about climate change, dependence on foreign oil, and growing household costs attributable to transportation. The federal government should require greater coordination among transportation and land use planning to promote more transportation choices for Americans, in response to the growing pressures of rising gas prices and congestion. One way to promote greater land use and transportation coordination is to direct greater transportation funding and decision-making powers to regional authorities, such as Metropolitan Planning Organizations and regional transit authorities. Such a move recognizes the growing needs of metropolitan regions to maintain aging urban infrastructure systems, the strong role of metropolitan regions in the nation's economy, and the opportunity within regions to provide viable public transit systems that maximize existing land uses.

A national infrastructure plan that responds to the three challenges above could provide an organizing framework for the actions at the federal level that are needed to leverage state, local, and private investments. This plan could take shape in a variety of ways, including through a commission-led process, as proposed in 2008 legislation to create a US commission on renewing and rebuilding America, or a civic-led initiative of national outreach and regional forums in collaboration with government and industry leaders. No matter the process, the end result should include the strategic priorities of the federal government to guide another generation of growth that can be shared by every community and region in the country.

Conclusion

This chapter opens with an account of the enormous challenges now facing the United States, including growing global competition, climate change, and the urgent need to rebuild economic opportunity in underperforming communities and regions across the country. As Americans mobilize to meet these challenges, they can be inspired by plans and actions taken by earlier generations of Americans, who time after time, when faced with even more daunting tests, met them with foresight and courage. Over the coming years, America will have the opportunity, under the leadership of the Obama administration, to take bold steps to invest in the nation's infrastructure.

Notes

1 According to the US Congressional Budget Office, since the mid-1980s, US public spending on infrastructure as a share of GDP ranges from 2.3 to 2.5 percent.
2 Adapted from Robert Fishman (2007) and correspondence with Roger Kennedy, historian.
3 The Strategic Vision Study process of the I-95 Corridor Coalition in the Northeast, Southeast, and Florida, and the coordination among regional agencies and ports in the Northern California Trade Corridor Coalition convened by the Bay Area Council, present two current examples of ad hoc collaborations.

References

AASHTO (American Association of State Highway and Transportation Officials) (2007) *Transportation Invest in Our Future: A New Vision for the Twenty-First Century*, Washington, DC: AASHTO.

Allbee, S. (2005a) "America's Pathway to Sustainable Water and Wastewater Systems," *Water Asset Management International*, 1(1): 9–14.

—— (2005b) "Finding a Pathway for Sustainable Water and Wastewater," *Underground Infrastructure Management*, November/December.

Appleton, A. (2007) "Non Structural Alternatives to Capital Needs," *Proceedings of the Water Environment Federation*, WEFTEC 2005: Session 81 through Session 90: 7196–7202(7).

ARTBA (American Road and Transportation Builders Association) (2007) "A New Vision and Mission for America's Federal Surface Transportation Program: ARTBA Recommendations for SAFTETEA-LU Reauthorization," Executive Summary. Online at www.artba.org/pdf/SAFETEA-LU_Recommendations_Exec_1107.pdf (accessed December 2007).

ASCE (American Society of Civil Engineers) (2005) *Report Card for America's Infrastructure*. Online at www.asce.org/reportcard/2005/page.cfm?id=203 (accessed December 2007).

Congressional Budget Office (2007) *Trends in Public Spending on Transportation and Water Infrastructure, 1956 to 2004*. Online at www.cbo.gov/ftpdocs/85xx/doc8517/08-08-Infrastructure.pdf (accessed August 2007).

Downey III, M.L. (2007) *Legislative Considerations for Long Term Policy Change: The National Roundtable on Surface Transportation: Discussion Papers and*

Summary, New York: Regional Plan Association and the Lincoln Institute of Land Policy.

Fishman, R. (2007) "1808–1908–2008: National Planning for America," in Regional Plan Association *America 2050: Framing Papers for the Rockefeller Foundation Global Urban Summit*, New York: Regional Plan Association.

Friedman, T.L. (2008) "Dumb as We Wanna Be," *The New York Times*, April 30, 2008.

Howes, J. (2007) "Taking Environmental Protection to the Next Level: An Assessment of the US Environmental Services Delivery System," *A Report by a Panel of the National Academy of Public Administration for the United States Environmental Protection Agency*. Washington, DC: The National Academy of Public Administration.

IPCC (Intergovernmental Panel on Climate Change) (2007) *Climate Change 2007: AR4 Synthesis Report*. Online at www.ipcc.ch/pdf/assessment-report/ar4/syr/ar4_syr.pdf (accessed November 17, 2007).

Kavinoky, J. (2008) "Statement of the US Chamber of Commerce on Surface Transportation Needs and Funding to the US," *House of Representatives Committee on Budget*, Washington, DC: The US Chamber of Commerce.

Mahone, A. (2007) "Policy Options for Climate Change Mitigation, Adaption and Energy Security," *United States Regional Plan Association: Requested for America 2050 Conference*, Bellagio, Italy.

Milwaukee Journal Sentinel (2008) "Watershed Trust can be a Force for Regional Unity," Editorial, *Milwaukee Journal Sentinel*, May 3.

National Surface Transportation Policy and Revenue Study Commission (2007) *Transportation for Tomorrow*. Online at www.transportationfortomorrow.org/ (accessed January 15, 2007).

O'Keefe, S. et al. (2007) "Prioritizing America's Water Resources Investments: Budget Reform for Civil Works Construction Projects at the US Army Corps of Engineers," *A Report by a Panel of the National Academy of Public Administration for the US Congress and the US Army Corps of Engineers*, February, Washington, DC: The National Academy of Public Administration.

Otto, B. (2008) "Floods and Drought, Building Community Resilience in the Face of Climate Change," *Mayors Water Council Newsletter, US Council of Mayors*, Winter.

Puentes, R. (2008) "Beginning Again: A Metropolitan Transportation Vision for the Twenty-First Century," *Statement before Committee on Transportation and Infrastructure, US House of Representatives*, April 9, Washington, DC: The Brookings Institution.

Reuters, I. (2007) *India Needs More Infrastructure Spending – fin min*. Online at http://in.reuters.com/article/businessNews/idINIndia-29371020070906 (accessed September 2, 2007).

Sassen, S. (2007) *Megaregions: Benefits beyond Sharing Trains and Parking Lots? The Economic Geography of Megaregions*, The Policy Research Institute for the Region, Woodrow Wilson School of Public and International Affairs, Princeton University. Online at www.princeton.edu/prior/publicatons/docs/megaregions.pdf (accessed February 9, 2007).

Singer, S.R. (2007) "Beijing's Thirty-year Boom," *National Interest*. Online at www.nationalinterest.org/Article.aspx?id=14942 (accessed July 12, 2007).

United States Congress (2007) "Energy Independence and Security Act of 2007." Online at http://thomas.loc.gov/cgi-bin/bdquery/z?d110:H.R.6: (accessed April 2, 2010).

United States Department of Energy Office of Electricity Delivery and Energy Reliability (2008) *Electricity 101*. Online at www.oe.energy.gov/information_center/faq.htm#ppl1 (accessed April 22, 2008).
United States Department of Energy Office of Electric Transmission and Distribution (2004) National Electric Delivery Technologies Roadmap: Transforming the Grid to Revolutionize Electric Power in North America, Washington DC: US Dept. of Energy Office of Electric Transmission and Distribution.
United States Environmental Protection Agency (2002) *The Clean Water and Drinking Water Infrastructure Gap Analysis, US EPA Office of Water*. Online at www.epa.gov/safewater (accessed September 2002).
Weinshel, K. (2008) "State: Regulators Predict Smart Grid R&D Funding for BPL", *UPLC PowerLine*, January 16, 2008. Online at www.uplc.org/news/2008/01/16/uplc-powerline-1-16-08 (accessed May 1, 2008).
Zheng, A. (2007) *Smart Grid Newsletter: Grid Automation*. Online at www.smartgridnews.com/artman/publish/Grid_Automation/Next-Generation_Grid_Visualization_251.html (accessed April 24, 2008).

7 Mega-city regional cooperation in the United States and Western Europe

A comparative perspective

Linda McCarthy

Introduction

Globalization and recent changes in governance associated with the shift toward neoliberal policies have impacted local governments on both sides of the Atlantic. Many feel forced to become ever more entrepreneurial in pursuit of jobs and revenues, and increasingly pro-business in terms of their expenditures. On the one hand, urban sprawl beyond the boundaries of individual cities means that metropolitan regions and the larger mega-city regions (MCRs) have assumed increased importance as national engines of growth in a global economy. On the other hand, the long-established autonomy of US local governments and recent national decentralization trends in Western Europe come with politically fragmented landscapes. Within MCRs, there may be an inefficient lack of coordination of policy and spending on economic development such as transportation infrastructure, and wasteful competition among localities using subsidies to attract inward investment by foreign companies.

Despite pressing cross-cutting issues, localities within an MCR may have little incentive to consider the impact of their own policies on adjacent jurisdictions or to modify their policies to benefit their entire MCR. Paradoxically, there is recognition of the need for localities to cooperate in order to be competitive in the global economy. Yet, cooperation is not easy, and an MCR's enormous size and complexity add additional obstacles.

Mega-city regions

A consistent MCR definition based on size is needed for international comparative analysis. This chapter focuses on four MCRs—London, Paris, New York, and Los Angeles—that correspond in enormous size (at least 10 million in population) to MCRs on a global scale. Although not commonly used in the US, the term "mega-city region" is increasingly used in Europe, especially within the context of the European Spatial Development Perspective (ESDP) of the European Union (EU). Applied within a European context, however, the size of some MCRs—such as Dublin or Rhine-Main—can be quite small (see Peter Hall's chapter in this volume).

To further control for differences, these four MCRs have functional polycentricity, involving a dominant core with strong intra-MCR functional linkages. Any finding of limited cooperation will not be due to morphological polycentricity, involving multiple smaller cities with weak intra-MCR functional linkages where a shared identity is less likely.

MCR cooperation

City regional cooperation is also referred to as "metropolitan regionalism" (Brenner 2002) or "competitive city regionalism" (Cisneros 1995; McCarthy 2003). This generally involves local governments, often with the involvement of higher tiers of the state, cooperating in networks that can include private interests and nonprofit agencies, that focus their economic development efforts to benefit their metropolitan region *as a whole*. This is "governance," rather than "government" or "inter-firm alliances." The private sector is a target *and* a participant; new businesses are targeted, while existing companies participate in the cooperative networks.

The focus can be exogenous and/or endogenous—for example, joint public–private promotional efforts on a metropolitan scale to attract new investment or shared investments in communications infrastructure across a metropolitan region to promote favorable conditions for existing and start-up companies. The assumption is that cooperation can address some of the shortcomings of uncoordinated efforts and wasteful competition within an MCR. Benefits include: economies of scale from pooling resources for more cost-effective spending; the potential to better market a metropolitan region for investment and reduce zero-sum competition; helping a metropolitan region mobilize its strengths to address socio-economic divisions that can weaken its chances of success; and finding a profitable niche in the global economy (Cisneros 1995).

Less formal cooperation is often initiated from the bottom-up by local participants and can be short-term or "weak," such as efforts to improve competitiveness through local government best practice exchange. More formal cooperation can be top-down or bottom-up initiated, and involve "strong" efforts to address political fragmentation, including establishing a metropolitan government or tax-base sharing (Walker 1987).

Most of the urban literature focuses not at the scale of MCRs such as London or New York but at the scale of smaller metropolitan regions such as Minneapolis-St. Paul and Montpellier. The Twin Cities Metropolitan Council was established in 1967 by the state legislature to address problems of urban growth. The governor appoints the public, private, and nonprofit council members. The Metropolitan Council coordinates development using review powers over local government policy in areas such as sewers and open space protection. It plans for metropolitan infrastructure, such as airports and transit. Its binding development plan includes a growth boundary to encourage compact development and protect environmental quality and agricultural areas. Local government tax-base sharing reduces fiscal disparities between growing and declining parts of the metropolitan region.

The Montpellier Languedoc Roussillon Technopole was established in 1985 as a cooperative network of politicians, entrepreneurs, researchers, and academics to promote company start-ups, introduce new technologies to existing companies, and assist new businesses locating in the area. In 1999, it came under the umbrella of the Montpellier Agglomération of thirty-one municipalities with an elected council. Structures—such as public commissions and neighborhood committees—allow citizens to be informed and voice their opinions. Technical areas promoted include medicine and biotechnology, computer technology, and agro-industrial research. Cooperative efforts include joint projects between public universities and private businesses. The region has received EU funding for major development projects and science and technology programs; it is permitted by the EU to provide public incentives.

In the last few centuries, the US and Europe have been home to some of the largest and fastest-growing urban agglomerations in the world. An examination of urban governance and strategic planning in their largest MCRs may offer useful insights because their slower growth and greater prosperity now may allow them to enjoy more policy options than faster-growing MCRs in less developed parts of the world.

This chapter reviews the economic development and public policy literature to investigate a number of research questions. First, what is the state of the theoretical research on city regional cooperation and how useful is it for developing a conceptual framework for examining MCR cooperation in comparative perspective? Second, what are the main cooperative efforts and institutional arrangements in the London, Paris, New York, and Los Angeles MCRs? Third, what are some theoretical insights and policy implications concerning cooperation that may be useful not only for MCRs in developed countries but also in other parts of the world, despite differences in political, economic, demographic, and other contexts?

Conceptual framework for understanding MCR cooperation

The recent changes in metropolitan governance have been accompanied by a growing number of theories and interdisciplinary debates. This section is a brief review of four main bodies of literature that provide insights into cooperation in MCRs that can inform a conceptual framework for this chapter.

The first body of literature considers the city region and metropolitan cooperation from an economic perspective within the context of globalization. Allen Scott (2001: 4–5) argues that global city regions "are coming to function as the basic motors of the global economy, a proposition that points [. . .] to the further important notion that globalization and city-region development are but two facets of a single-integrated reality." Research on "competitive city regionalism" (Cisneros 1995; McCarthy 2003), although not necessarily applying to the scale of MCRs, is relevant because it focuses on cooperation to promote the competitive advantage of city regions based on the quality of internal attributes such as workers and infrastructure.

Although research on world cities has traditionally measured these kinds of attributes, recent literature has emphasized that city regions are part of a global network (Hall 2001). This external focus highlights the importance of global economic restructuring with city regions as competitors for international investment. This focus has been criticized, however, for reifying city regions as independent economic units within the global economy.

A second body of literature takes a more political perspective. Research on the "rescaling" of the national state argues that territorial competitiveness is increasingly promoted at the scale of the city region (Brenner 2003). It is argued that the changing relations between different tiers of government occur in conjunction with what some see as the "hollowing out" of the national state (Ohmae 1995). A criticism of this literature is that it has neglected the continued, if reconfigured, role of national government.

Brenner (2002) argues that "metropolitan regionalism" is not so much a coherent state-led political project as an incoherent response to economic restructuring that is rooted in the crisis of Fordism. This disregards earlier proactive cooperation, as in Minneapolis-St. Paul due to concern for the environment and quality of life amid fears of septic tank failure and groundwater contamination with uncontrolled sprawl. The political literature, however, facilitates an understanding of the lack of cooperation and conflict among multiple participants, horizontally within MCRs and vertically among different tiers of government.

A third body of literature addresses social issues related to citizenship and democracy. This work includes research on global cities by sociologists concerned with social inequalities (Abu-Lughod 1999; Sassen 2000). This research has been augmented recently with a more explicit spatial component. Jonas and Ward (2007: 174), for example, call for more emphasis on the social relations of production, consumption, and quality of life within and between city regions given uneven development and redistributional concerns. They argue that "age-old contradictions of spatial growth in capitalism come to rest in and through the territorial structures of city-regions." They point to the emergence of social movements that contest the economic emphasis of metropolitan cooperation. This literature highlights the importance of addressing social inequalities and uneven development as part of cooperative efforts that can benefit an entire MCR.

A fourth recent body of literature is concerned with cooperation and environmental quality in city regions (see, for example, Keil 1998). This literature remains under-theorized and could be better integrated into existing economic, political, and social conceptualizations of MCRs, given its cross-cutting nature:

environmental degradation in MCRs comes with negative social effects, involving high economic costs and requiring political intervention. Environmental and economic goals overlap, because the environment cannot be divorced from economic context; increasingly, environmental quality is seen as an important attribute in a city region's competitive arsenal. Environmental quality, as an issue, crosses jurisdictional boundaries and needs to be addressed through cooperative efforts.

Although these literatures have their own viewpoints and assumptions, a comprehensive conceptualization of MCR cooperation must incorporate all four perspectives. Economically, with the importance of city regions as engines of growth, cooperation can be understood as a potentially effective strategy for jurisdictions within an MCR; the goal is to retain and attract businesses in order to promote economic growth, including job creation, in a context where the driving force is competitiveness in the global economy. Moreover, efficiencies from economies of scale in public infrastructure and services provision can be achieved through better coordination.

Politically, with territorial competitiveness increasingly considered at the city region scale, fragmentation is a factor in understanding why cooperation—from informal governance to formal government efforts—is difficult, and also why there may be internal conflict among multiple public, private, and nonprofit actors both horizontally within an MCR and vertically from local through higher tiers of government.

Socially, the notion of equity, for altruistic or self-interested reasons, underlies efforts to jointly address inequality. The interconnected nature of an MCR and the potential for negative spillovers mean that improving the quality of life of residents in poorer parts of an MCR through education, training, and so on can also improve the quality of life of the entire MCR. Environmentally, the fact that environmental problems do not respect fragmented political boundaries explains interest in joint initiatives.

Bids for the 2012 Olympics

Before using this conceptual framework to consider the main cooperative efforts and institutional arrangements in the London, Paris, New York, and Los Angeles MCRs, it is useful to show how it can inform an examination of cooperation involved in the bids for the 2012 Olympic Games. That London was successful is less informative than an examination of the institutional frameworks, funding arrangements, and economic, social, and environmental goals that allows the differences between these four MCRs to be highlighted.

London's bid was initiated by the British Olympic Association, who enlisted the Mayor, along with national government (as the initially reluctant major funder) after private-sector lobbying. The "Back the Bid" campaign to get the public on board featured an advertising campaign on billboards and buses. The International Olympic Committee (IOC) poll indicated that only 68 percent of Londoners supported the bid. The low percentage reflected concerns about cost, including a £20 annual tax rise to fund the Olympics. London Citizens, a coalition of sixty community groups, made demands, including a living wage for Olympic workers, which were met in order to create a united front for the IOC's visit. The Prime Minister called on the public to show their "passion" for a London Olympics and lobbied in Singapore during the build-up to the decision. The Lower Lea Valley, the location for the new Olympic Park, will be the largest urban park created in Europe in over 150 years. Economic, social, and environmental goals include increased tourism, regenerating brownfield sites, revitalization associated with transportation enhancements, over 4,000 new homes in the Olympic Village, and a wildlife corridor.

The Paris bid was handled by a public interest group chaired by the Mayor, and comprising the French National Olympic Committee, the City of Paris, the Île-de-France region, and the French Ministry of Youth and Sports. Funding was to come from local, regional, and national government, which guaranteed any shortfall. A campaign to get the public on board included decorating monuments with the Paris 2012 logo and closing the Champs Élysées for a sports festival/street party. The IOC opinion poll showed an overwhelming majority—85 percent—of Parisians supported the bid. One reason may have been that France had hosted the 1998 World Cup and the new stadium could be used to keep costs down. Economic, social, and environmental goals included increased tourism, regenerating an industrial wasteland, and new housing at the Olympic Village.

New York's bid was private-sector initiated and led by a private nonprofit organization. The Mayor and State Governor were firmly on board. Funding was to come from private sources, with public funds for needed infrastructure. The IOC poll indicated that a mere 59 percent of New Yorkers supported the bid. Midtown West neighborhood groups objected to a new stadium that they feared would bring congestion and unwanted neighborhood change. The New York Jets were putting up $800 million, the City $300 million for infrastructure, and the State $300 million for a retractable stadium roof. Strong private support was evident when a State board failed to allocate funds for the roof, potentially jeopardizing the bid; the New York Mets stepped in and offered their new baseball stadium in Queens, with $250 million for temporary conversion of the stadium covered by private and public dollars. The economic, social, and environmental goals included revitalizing the East River waterfront, which would include new housing in the Olympic Village.

The Los Angeles bid was initiated by a civic organization, with the support of the City, and was based on the 1984 Games concept of being completely privately funded. This bid was placed second after New York, which became the US bid city. As Los Angeles was eliminated so early, the IOC did not commission an opinion poll, but one by Wirthlin Worldwide found that 85 percent of Los Angelinos supported the bid. The private funding, an expected financial surplus, and limited new infrastructure investments were factors in this strong support. Economic benefits included increased tourism.

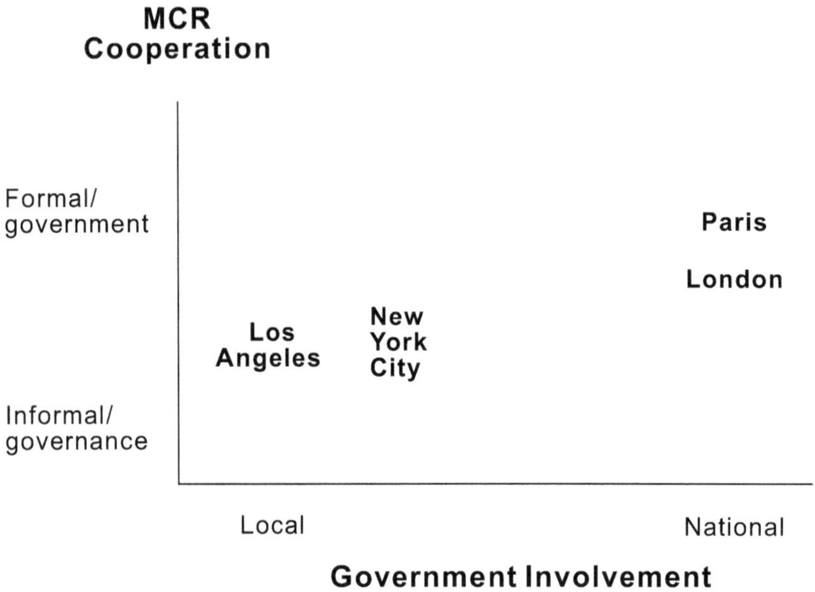

Figure 7.1 2012 Olympic bids

Social and environmental benefits were few due to the reuse of many venues from the 1984 Olympics and the use of new facilities built since 1984. No housing construction was included because of plans to use the recently expanded residence halls at the University of California, Los Angeles (UCLA) and the University of Southern California (USC) as Olympic Village accommodation.

Figure 7.1 shows a continuum for government involvement—from national to local—and another for cooperation—from formal/government to informal/governance. The four MCRs can be located on this graph based on the 2012 Olympic bids of their core cities. Reflecting the stronger hand of national government in Europe, the Paris bid was the most publicly driven and funded, slightly ahead of London. In contrast, reflecting more active private-sector participation in the US, the Los Angeles bid was the most locally and private-sector driven and funded, followed closely by New York.

United States and Western European MCR cooperation

This section uses my conceptual framework to survey the main cooperative efforts and institutional arrangements in the London, Paris, New York, and Los Angeles MCRs. Having considered the EU context for Europe and the national context for the US, each MCR is examined as systematically as possible with reference to: political fragmentation; metropolitan governance and strategic planning; and competition and cooperation.

Western Europe

Competition within the EU to attract exogenous investment using public subsidies is generally prohibited because they provide unfair advantage to companies receiving the subsidies. Exceptions include subsidies for promoting economic development in areas with abnormally low standards of living or serious underemployment. Subsidies can be for small companies, industrial restructuring, retraining, or job creation.

The ESDP locates London and Paris within the strong core area with a high standard of living and well-developed infrastructure. Although playing important roles in the ESDP's goal of polycentric and balanced spatial development, and containing pockets of severe deprivation, London and Paris receive limited EU funds compared to needier regions. At the same time, this discussion fits well within a theme of this book—how European MCRs are evolving and strategic intervention redefined in an attempt to enable the integration of urban qualities in a multilevel governance environment.

London

Politically, the London MCR is fragmented, being composed of the City of London and thirty-two boroughs (comprising Greater London), numerous single-purpose agencies including urban development corporations (UDCs), and those jurisdictions extending outside the Green Belt but within the built-up area of about 19 million people.

Important for metropolitan governance was the national government's establishment in 2000 of the position of Mayor and the Greater London Authority (GLA). The Mayor sets the budgets for the GLA, Transport for London, the London Development Agency (LDA), the Metropolitan Police, and London's fire services, and has other responsibilities, including planning and economic development, housing, and the environment. The national government retains considerable control over much of the Mayor's budget. Moreover, the Mayor does not have complete strategic planning or budgetary control over public services across the entire Greater London area because he shares responsibility for many functions with other national and local agencies.

The Mayor leads the preparation of the London Plan, a strategic plan for Greater London, which involves responding to business demands including business lobbies that have a vested interest in the contents of the Plan. The jurisdiction of the GLA excludes an extensive area of the MCR surrounding Greater London, which is not included in the London Plan. On paper, the London Plan acknowledges Greater London's position and interactions within the larger MCR, support for polycentric development across the MCR, and acceptance of the national government's proposed growth areas to spatially balance new development, such as the Thames Gateway. In practice, until recently, there was limited cooperation and coordination across Greater London (John et al. 2005), even excluding those areas outside the Green Belt.

In an effort to promote competition and coordination, the national government established three regional development agencies (RDAs) for the London MCR in 1997. The RDAs are run by business-led boards, which contain corporate leaders, public officials, and nonprofit representatives. The RDA's twin goals work against each other—promoting economic competitiveness trumps addressing political fragmentation. As reflected in their policy documents, economic competitiveness drives local and RDA agendas more than social equity or environmental concerns (Herrschel and Newman 2002).

As an RDA, the LDA is mainly concerned with the competitiveness of its own territory—Greater London—rather than that of the MCR. Its goals are economic, social, and environmental: business growth, socio-economic inclusion and diversity, and sustainable development, respectively. There is no requirement that these goals are coordinated with the other two RDAs and no cooperative mechanism for them to do so. The LDA partners with businesses, nonprofits, and the community. Citizen participation mainly involves voting for the Mayor and GLA Assembly members; citizens also attend meetings and the biannual question time, and submit petitions.

The Mayor established the London European Forum in 2001 as a high-level advisory body of public, private, and nonprofit participants. London already had a European Office in Brussels funded by the GLA and other public agencies. The London European Forum is intended to allow its members—from Greater London—to share information and keep abreast of issues emerging in Brussels in order to lobby the EU in a more coordinated manner.

London Councils is an association comprising the thirty-two boroughs and City of London, Metropolitan Police Authority, and London Fire and Emergency Planning Authority. It lobbies higher tiers of government, and provides a forum for information exchange.

London and Paris are members of Eurocities, the network of major European cities that is partially funded by the EU, and collaborates in best practice and information exchange; they are also members of EUKN—the EU's European Urban Knowledge Network for sharing information about tackling urban issues—and of METREX—the Network of European Metropolitan Regions and Areas, which receives EU funding. This cooperation excludes the area outside Greater London or Greater Paris. Only parts of London such as Medford and parts of Paris such as Nanterre are active in the forum for information exchange of the EU's URBACT, the European program for sustainable development. The "weak" nature of this cooperation is evident in the LDA's list of achievements, which includes how it "attracted 20 percent more projects than its closest rival, Paris" and had maintained its position "as Europe's top location for business" in 2005.

Ironically, political fragmentation within the MCR is reflected in a 2008 bottom-up cooperative initiative between London and New York: the Innovative Exchange Programme. Signed only by the two mayors, it involves officials from both cities sharing ideas on issues such as law and order.

According to New York's Mayor, "New York and London have a special relationship as two of the world's greatest cities. We not only compete with each other, we learn from each other." This more informal cooperation does not include the surrounding jurisdictions of either MCR.

An example of cooperation involving multiple participants is the Thames Gateway to the east of London. This area was selected by the national government as a project with economic, social, and environmental goals: to promote economic development to improve London's status as a dominant world city, provide affordable housing, and redevelop the UK's largest brownfield site. It has received EU funding.

Promoted by national government as a coherent geographical entity united by a single strategic vision, it appears ripe for cooperative efforts. The Thames Gateway's website shows a cooperative organizational structure; beneath the national department responsible for coordinating the project are public agencies, private businesses, and nonprofit organizations. The challenge for cooperation is that it straddles London's three RDAs, two county councils, and fourteen local governments, as well as the GLA, nationally established UDCs, and the national regeneration agency, all of which have different management structures and public, private, and community participants (John et al. 2005).

In addition, the Thames Gateway is split into three zones corresponding to the three RDA areas. The two areas in the LDA's jurisdiction are north and south of the Thames. In the north, the Thames Gateway London Partnership is a public–private partnership comprising twelve local governments with business interests such as those in Canary Wharf; the "delivery vehicle" is the London Thames Gateway Development Corporation, a UDC. The idea is to build on earlier accomplishments of the London Docklands Development Corporation (LDDC), a UDC for Canary Wharf. There, the creation of an international financial services district to compete for global business was initially envisioned by private interests, who persuaded the LDDC and national government to support the project.

Overall, the national government is the dominant actor; it maintains strong top-down control over policy and strategic planning affecting the MCR and projects such as the Thames Gateway. Although some responsibility has been devolved to lower levels of government, this has not come with significantly increased local power, such as autonomy to raise local taxes as in the US. Lower tiers of government take social and environmental issues such as affordable housing and sustainable development seriously; financial support from national government, however, requires that the national rather than any local agenda is followed.

In fact, national decentralization policies, whether intentional or not, created a politically fragmented situation where there is a lack of governance at the MCR scale (see Peter Hall's chapter in this volume). Until 2000, strategic planning was organized on a formal basis through the South East Regional

Planning Conference (SERPLAN) comprising representatives from the county planning departments who advised national government on regional issues, and RPG9, the national government's regional planning guidance for the South East. Since 2000, strategic planning has been done on an informal basis; the national government negotiates with each planning authority, which encourages a local rather than cooperative MCR perspective. Although the ideal of cooperation within the MCR is accepted in many local planning documents, political fragmentation means that it has fallen to national government to assume responsibility for strategic policy and planning for the MCR; that same political fragmentation, however, makes this task incredibly difficult (Herrschel and Newman 2002; Salet et al. 2003; John et al. 2005).

Paris

Politically, the Paris MCR is fragmented, being composed of the Île-de-France regional council, which coincides with eight départements and over 1,300 municipalities and eight counties, as well as single-purpose agencies such as the Institute for Urban Planning and Development of the Île-de-France region (IAURIF), and those jurisdictions outside the Île-de-France but part of an MCR of almost 16 million people.

Compared to the UK, the French government has decentralized responsibility but also power to lower tiers of government. In 1982, the national government established the regions as a local tier of government. The Île-de-France region corresponds with Greater Paris but does not include the entire MCR. National government sought further coordination by establishing single-function directorates for each region, including one for environmental protection (DIRENS) and one for public works (DREIF).

The Île-de-France region approved a master plan in 2007. This long-range strategic framework coordinates a range of public policies for housing, transportation, social inequalities, open space, and urban sprawl. Unlike the master plans of the other French regions, the Île-de-France plan includes a prescriptive land use document with which the municipal master plans must comply (Fouchier 2008).

In 2000, a regional development agency for Île-de-France—Agence Régionale de Développement (ARD)—was established at the initiative of public and private agencies (Île-de-France regional council, Paris Chamber of Commerce, and Regional Chamber of Commerce and Industry). Like the LDA, ARD is mainly concerned about the competitiveness of its own territory—Île-de-France—rather than that of the entire MCR. Despite acknowledging that sustainable development should be integrated throughout its efforts, ARD's primarily economic goals—marketing the region abroad and attracting foreign investors—are not required to be coordinated with jurisdictions adjacent to Île-de-France. In contrast, the two transit agencies operating the city center subway and regional rail lines coordinate their systems for a seamless fare structure for passengers across the MCR.

The French decentralization laws encourage local cooperation. The 1999 Chevènement Act established a framework for groups of municipalities to overcome political fragmentation and cooperate as a community of agglomeration in policy and services such as transportation infrastructure. Business taxes and additional revenues, such as the household waste removal tax, are collected communally, as is a state grant; this is doubled for cooperating jurisdictions—the real financial incentive of the Act (Négrier 2006). Instead of coordinated results across the entire MCR, the Act instead has allowed narrow sub-regional interests to emerge, including suburban groupings pitted against central Paris (Subra and Newman 2008).

Some of the eight Île-de-France départements have their own competitive strategies. The most prosperous département after Paris, Hauts-de-Seine, which includes La Défense business district, actively attracts international company headquarters. Val d'Oise's economic development agency has "embassies" in several countries; it markets its economy, quality of life, and location relative to central Paris and nearby Roissy-Charles de Gaulle Airport (Lefèvre 2003).

Cooperation involving multiple public and private interests includes the local governments of the Community of Agglomeration of the Plateau de Saclay, and the Strategic Innovation Council, an independent French think tank of public and private members. This group lobbies for national and EU funds for investments in infrastructure to support science and technology in the Saclay *pôle d'excellence* to help Paris better compete with other MCRs such as London and New York. The aim is to build upon major national investments in projects—such as the Soleil synchrotron electron accelerator and the Orphée neutron reactor—that have already achieved a concentration of cutting edge companies, universities, and research institutes.

Paris Rive Gauche is another cooperative effort. The original idea was to provide office space in the office-poor historic core as an eastern counterbalance to La Défense to allow Paris to compete for global business against other European capitals, including London. Following objections from a local community group, the project was modified to a mixed-use development with 50 percent affordable housing and 25 acres of green space. This planned development zone of former industrial land and railway yards comprises a 321-acre business district containing office, university, and apartment buildings. Funding came from the City of Paris (57 percent), SNCF (French railways) (20 percent), which owns 80 percent of the land, RIVP (Paris building authority) (10 percent), the French government (5 percent), Île-de-France region (5 percent), and private sources (3 percent). SEMAPA, a public-private company, oversees the project, provides the infrastructure, and sells land parcels.

Decentralization has brought planning closer to the citizens; it allowed resident associations to become part of the planning process (Subra and Newman 2008). At the same time, although the role of national government has been diminished, it remains the strongest actor (Lefèvre 2003). Moreover, despite decentralization laws, political fragmentation and a lack of governance at the MCR scale are similar to that found in London due to the mismatch between administrative boundaries and areal extent of the MCR.

United States

The federal government is not active in discouraging wasteful competition involving public subsidies. US localities have considerable autonomy over economic development and control a major revenue stream—property taxes. In addition to job creation, higher property taxes are a goal of local competition for private investment.

Each individual state, however, discourages wasteful competition among their own jurisdictions when the outcome is likely zero sum for the state. The states decide which locality to support, and play a significant role in assembling incentive packages for high-profile companies, because a single locality cannot afford to do this alone (McCarthy 2003).

In contrast to Europe, discussions about cooperation in the US focus more on governance than on new metropolitan tiers of government. Downs (1994: 182) concluded that introducing metropolitan governments to act as "regional policy czars" was unachievable due to resistance from all tiers of government concerned about loss of authority and revenues.

There are, however, federal laws requiring coordination of policies within metropolitan regions; they are limited and apply only to transportation, air quality, higher education, and health facility planning. The 2005 Safe, Accountable, Flexible, and Efficient Transportation Equity Act (SAFETEA) promotes cooperation among localities because it requires integrated transportation at the metropolitan scale. SAFETEA requires and helps fund metropolitan planning organizations (MPOs)—councils of governments (COGs) comprising elected officials from each municipality, or regional planning councils (RPCs) staffed by professional planning officials—with responsibility for devising metropolitan transportation plans and deciding which local transportation projects receive federal funding. This recent federal support for metropolitan planning is not comprehensive because each federal agency promotes regional coordination for its own programs rather than in an interconnected way across all federal agencies (Pastor et al. 2000).

Within the context of a theme of this book—how traditional federal states see the promise of major policies and development initiatives finally moving ahead to herald more strategic intervention at national and regional scales—the Obama administration may herald a new phase of federal involvement in urban policy. The new White House Office of Urban Affairs seeks to achieve more coordinated federal programs that produce more effective and efficient outcomes for cities. This is especially necessary because of the current economic downturn and federal budget deficit.

The approach is at the scale of the metropolitan region—disregarding traditional jurisdictional boundaries and establishing policy that reflects the interconnections between cities and suburbs. This integrated approach is intended to facilitate enhanced economic competitiveness, environmental sustainability, and social equity. The goal is to benefit entire metropolitan regions through federal funding that focuses on core infrastructural investments, such as the construction of sustainable power, transportation, water, and telecommunications networks. This federal urban policy agenda involves a pragmatic pro-business perspective grounded in encouraging job creation through partnering with the private sector, including support for businesses in regional clusters of innovation and next-generation industries

This more activist and coordinated federal government may be able to better support local government decision-making than previously. It may benefit entire metropolitan regions and, more crucially, the varied territorial constituencies within them who are engaged in building partnerships to leverage new federal capacities and resources as they become available.

New York

Politically, the New York MCR is fragmented, being composed of sixty municipalities, thirty-seven counties, and numerous single-purpose agencies across four states: New York, New Jersey, Connecticut, and Pennsylvania. The population of the New York Consolidated Metropolitan Statistical Area (CMSA) is over 21 million. A CMSA is a very large area comprising multiple communities around a large population center sharing strong internal economic and social linkages including commuting patterns.

The need for coordination across the MCR has long been recognized. The 1996 (third) regional strategic plan for the then thirty-one-county MCR by the Regional Plan Association (RPA) continues bottom-up efforts since the first plan in 1929. The 1996 plan highlighted the dilemma of maintaining competitiveness and growth in a global economy while also achieving social equity and environmental quality. The RPA acknowledges that, as an independent nonprofit group, mobilizing the public, private, and nonprofit sectors to implement the plan is difficult. The first plan's proposal to expand the regional rail system and the second plan's warnings against continued sprawl went largely unheeded (see Robert Yaro's chapter in this volume).

In response to federal laws, such as SAFETEA, the New York Metropolitan Transportation Council (NYMTC) was designated as the MPO for New York City, Long Island, and the lower Hudson River Valley. The NYMTC lobbies for federal transportation funding and is a forum for best practice exchange. There is a lack of coordination, however, because there are multiple MPOs in the multi-state MCR. Unlike the GLA and Île-de-France region, however, the MPOs as a group cover the entire MCR.

The MCR's transportation system is not coordinated across state lines. In New York, the state legislature set up the Metropolitan Transportation

Authority to oversee transportation only from New York City to Long Island. New York City and the seven counties around it finance their mass transit network through a combined sales tax surcharge.

The most common cooperative efforts are arrangements such as joint powers agreements, or special districts in which two or more municipalities agree to co-finance services such as emergency response. A larger version of the special district is a regional district or authority. A long-term—since 1921—example is the Port Authority of New York and New Jersey.

The Port Authority was an initiative by the states of New York and New Jersey to move past conflict over use of a port straddling both states. The board of this public agency comprises businesspeople appointed by each state governor. The Port Authority has considerable revenue-raising capacity (based on, for example, tolls on bridges between the two states) which raises accountability issues (Ross and Levine 2001).

Although not having jurisdiction for the entire MCR, the Port Authority could promote more coordination in its capacity as a regional planning agency; it focuses instead on infrastructure projects that generate revenue. It has attempted to encourage cooperation, such as devising a regional marketing strategy that localities refused to adopt. It convinced state and local governments to sign a regional "non-competition pact" in 1991 in which they promised not to offer incentives to attract companies away from neighboring localities. The pact fell apart when tested.

Competition has resulted in $215 million in public subsidies for Citigroup alone during the last eighteen years—state and local subsidies from New York and New Jersey essentially poached (the same) jobs from each other repeatedly. The high cost and wasteful nature of these subsidies would be better spent cooperatively to make the MCR more attractive by investing in improving the basic requirements of all businesses, including the quality of the labor force and infrastructure (Stecker and Steinberg 2007).

Competition is also vigorous within individual states; many jurisdictions operate their own economic development agencies. For example, the New York City Economic Development Corporation (NYCEDC) is a public-private economic development agency organized like a corporation but controlled by the mayor and other public officials. NYCEDC administers the New York City Industrial Development Agency (NYCIDA), which issues revenue bonds for economic development projects.

There is competition within individual jurisdictions. The State of New York passed a bill in 1996 that gave the five boroughs of New York City more opportunity to compete for new investment. The bill allocates a quarter of the NYCIDA's budget among the boroughs to fund their own economic development projects and provide tax relief for businesses. An editorial in the *New York Times* (1996) argued: "It's bad enough that City Hall feels it has to fight Stamford or Jersey City over luring business to New York. It makes no sense to let the Bronx, Brooklyn, Manhattan, Queens, and Staten Island join the fray."

Top-down initiated cooperation occurs within individual states. In 1968 the State of New Jersey created a regional body, the New Jersey Meadowlands Commission. It comprises public and private participants and acts as a zoning board to address environmental protection, economic development, and waste management. The fourteen localities across two counties agreed to tax-base sharing; this prevents individual municipalities from suffering financially when not allowed to develop protected land for commerce or industry. The Commission cooperated with environmentalists, businesses, mayors, and state and federal agencies to develop a master plan that emphasizes brownfield redevelopment, job creation, and wetland and open space preservation. In contrast, New York's Hudson River Valley Economic Development Corporation was established in a bottom-up initiative by eight counties, their cities, and a nonprofit environmental group; environmental and economic aspects involve marketing the brownfields to businesses.

Los Angeles

Politically, the Los Angeles MCR is fragmented, being composed of over 100 municipalities, as well as numerous single-purpose agencies in the five-county CMSA of more than 16 million people.

L.A. 2000 is a strategic plan produced by a public–private partnership that envisioned a "new regionalism" with economic and environmental issues balanced to achieve a more livable region. This report went further than the RPA's New York Plan; it argued for regional growth management and new regional government agencies to establish policy for land use, housing, transportation, water, pollution, and waste (Pastor et al. 2000; Ross and Levine 2001). This plan has seen the same lack of progress as the New York one for similar reasons.

The Southern California Association of Governments (SCAG) is the COG for the more than 150 municipalities in the six-county Los Angeles, Orange, San Bernardino, Riverside, Ventura, and Imperial area. As the only MPO for the Los Angeles MCR, there is more potential to coordinate transportation and planning than across the multiple MPOs in the New York MCR. SCAG's jurisdiction covers the entire MCR. Its members lobby for federal transportation funding. Elected officials from SCAG jurisdictions promote cooperation in order to meet federal regulations, including the Clean Air Act; SCAG works with the South Coast Air Quality Management District (AQMD), a special district responsible for air quality in the four-county LA basin.

A criticism of SCAG, and all COGs, is that its advisory status makes implementing its regional plans difficult. As a voluntary association, it has been criticized for steering clear of controversial issues out of concern that some local governments may leave. In the past, Orange and Ventura Counties unsuccessfully sought federal legislation to allow them to form their own MPO. Cultivating an MCR perspective is difficult because individual elected officials in SCAG are accountable to local rather than regional constituencies (Pastor et al. 2000; Ross and Levine 2001).

The metropolitan regional districts and authorities in the LA MCR tend to cooperate over single services. For example, the state established the Los Angeles County Metropolitan Transportation Authority with a board comprising elected and appointed officials, and the Southern California Metropolitan Water District involving cooperation by twenty-six municipalities and water districts including parts of Los Angeles, Orange, Riverside, San Bernardino, and Ventura counties. Although these regional districts achieve some cooperation, they do not cover the entire MCR and their revenue-raising capacity raises the same accountability issues as for the Port Authority in New York.

Cooperation can involve service contracts where one jurisdiction purchases services from another as in the Lakewood Plan involving Los Angeles County providing services to local governments more cheaply than they could themselves (Ross and Levine 2001). This has facilitated political fragmentation, however; small jurisdictions that would otherwise have agreed to be annexed by the City of LA were able to incorporate as independent local governments because they could purchase their services from the County.

Municipalities such as LA, however, are members of economic development agencies that promote investment for their county. The Los Angeles County Development Corporation (LACEDC) is an economic development agency comprising public and private members including individual cities such as LA, banks, and power companies. Competition is vigorous within the MCR, however, with many jurisdictions, including the City of Los Angeles, also operating their own economic development agencies.

Bottom-up initiated cooperation in parts of the MCR includes the Inland Empire Economic Partnership, which comprises local government and business partners in the western part of San Bernardino and Riverside counties. These jurisdictions lack a "cohesive territorial identity" compared to their rivals, Los Angeles and Orange Counties. Cooperative strategies include fiscal and land use tools to compete for inward investment (Jonas and Pincetl 2006). The depth of cooperation is limited, however, because individual counties and larger municipalities also compete individually using their own economic development agencies.

The Gateway Cities Partnership comprises businesses, labor organizations, educational establishments, and public agencies, including twenty-seven municipalities in southeast Los Angeles County. The focus is economic revitalization involving brownfield redevelopment, worker training, and traffic management to improve goods movement (Jonas and Pincetl 2006)

The private sector and local communities can complicate cooperative efforts at the MCR scale. Downtown LA business support has been crucial for cooperation but has put greater emphasis on the needs of the central business district (CBD), despite the economic distress of poorer neighborhoods. Cooperation is difficult if there is a perception that the CBD is getting too much attention, or if the central city is too dominant or seeking to impose costs on surrounding municipalities (Gainsborough 2003).

Not unexpectedly, there is a lack of governance at the MCR scale in the US. A major feature compared to Western Europe, however, is the more limited involvement of higher tiers of government in actively promoting cooperation and coordination.

Summary and conclusions: some theoretical and policy implications

In reviewing the academic and public policy literature and planning documents, it is clear that a city regional perspective and cooperation are more evident on paper than on the ground in the London, Paris, New York, and Los Angeles MCRs. There is limited cooperation even in London and Paris, despite the EU being touted as a supportive context for cooperation (see, for example, Keating 2001). The EU does not actively promote cooperation within London or Paris because, despite pockets of deprivation, these MCRs are not in areas with low standards of living and poorly developed infrastructures. With its European perspective, the EU encourages inter-urban and inter-regional cooperation among countries more than intra-urban cooperation within metropolitan regions in the same country.

Table 7.1 shows the main similarities and differences in cooperative efforts and institutional frameworks within the context of the comprehensive picture of MCR cooperation that my conceptual framework allows. The importance of MCRs as engines of economic growth means that national government involvement is important in understanding the extent of cooperation. The UK government introduced a metropolitan tier of government for Greater London but maintains strong control, especially over finance; the French government established a metropolitan tier of government for Greater Paris but decentralized more power and resources; and the US government requires coordination through the MPOs in New York and Los Angeles. Despite different national contexts, political fragmentation is the major obstacle to MCR-wide cooperation. The MCRs do not have a single agency with the capacity and resources to act.

Although government entities are understandably major players, the private sector is an important participant, albeit to a greater extent in the US (as, for example, in the Olympic bids). Cooperation is often, but not exclusively, top-down initiated in Europe; likewise, it is often but not exclusively bottom-up initiated in the US (see Table 7.1).

Cooperation to promote exogenous development and attract inward investment is not done at the MCR scale. The LDA and ARD have allowed greater coordination, but competition continues due to the competitive efforts of the other RDAs in London and the départements in Paris (see Table 7.1). When the UK government established the RDAs to address fragmentation and promote competitiveness—two incompatible goals—the result was

Table 7.1 Competition and cooperation

	London	*Paris*	*New York*	*Los Angeles*
Political fragmentation	Yes	Yes	Yes	Yes
Metro governance and planning	GLA and plan with some teeth	Île-de-France region and plan with some teeth	MPOs and plan without teeth	MPO and plan without teeth
Competition and cooperation	No MCR-wide cooperation; LDA and, e.g., 2 other RDAs; London Councils	No MCR-wide cooperation; ARD and, e.g., 7 départs; Com. of Agglom.	No MCR-wide cooperation; no Gtr NY agency and indiv. compet.; Port Authority? NYCEDC?	No MCR-wide cooperation; no Gtr LA agency and indiv. compet.; Lakewood Plan?; LACEDC? etc.
Examples of cooperation	Canary Wharf; Thames Gateway	Saclay; Paris Rive Gauche	NJ Meadowlands; Hudson River Corp.	Gateway Cities; Inland Empire
Endogenous vs. exogenous	Endogenous and exogenous	Endogenous and exogenous	Endogenous and exogenous	Endogenous and exogenous
Focus	Economic, social, environmental	Economic, social, environmental	Economic, social, environmental	Economic, social, environmental
Major player(s)	National gov.	National; local gov.	State; local gov.	Local gov.

competition. When a national Act offered French communities the opportunity to cooperate to address fragmentation and achieve competitiveness—two compatible goals—the result was competition; local governments merely "shifted scales" such that groups of municipalities cooperated to compete against other groups of municipalities within the Paris MCR. Within the New York and Los Angeles MCRs, wasteful competition is endemic because localities have autonomy over economic development. In the absence of federal restrictions on wasteful competition in the US, companies pit localities within an MCR against each other to raise the incentive offers. But despite the attention in the literature to the economic goals of cooperation, cooperation also includes endogenous social and environmental efforts, albeit often in support of overall economic competitiveness goals in the MCRs on both sides of the Atlantic.

From a theoretical perspective, this examination of cooperation suggests that a conceptualization of metropolitan cooperation as involving informal, flexible, purpose-driven and inherently unstable institutional arrangements in response to increased competition amid globalization (Herrschel and Newman 2002) neglects more formal proactive cooperative efforts to address social or environmental concerns, as in the Hudson River Valley. Likewise, a conceptualization of "metropolitan regionalism" as not so much a coherent state-led political project as an incoherent response to economic restructuring rooted in the crisis of Fordism (Brenner 2002) is incomplete. As a crisis response, for example, we might expect more localities within metropolitan regions to overcome their differences and cooperate more than they do.

There is a disconnect between theory and practice. The notion of the "rescaling" of the national state whereby territorial competitiveness is increasingly promoted at the scale of the city region fails to operationalize the connections between large-scale restructuring and metropolitan cooperation, especially where cooperation is limited. Economic restructuring may establish the context for cooperation, but the participants within a metropolitan region, with or without higher tiers of government, ultimately determine if cooperation will occur; cooperation is only one response to a crisis.

Moreover, higher tiers of government can establish supportive contexts for cooperation, but cannot guarantee that local public and private participants will cooperate at all or as intended. The induced cooperation may merely be enough to meet national requirements or access funds for "weak" efforts such as best practice exchange.

A strength of the political literature is the spotlight on internal conflict. Conflict is not an issue, however, if localities are not even considering cooperating. A survey of US cities by Kincaid (1997) found that, despite increased global competition, city officials continued to view localities within their own metropolitan region as their primary competitors, ahead of overseas localities. The greatest competition is usually between cities in the same metropolitan region (Goetz and Kayser 1993; McCarthy 2004).

The "rescaling" literature tends to define the metropolitan region rather casually. Brenner (2003), for example, talks about Greater London rather than the functional urban region. Neglecting those jurisdictions outside the Green Belt and the more complicated spatial pattern of the entire MCR has theoretical implications for notions of "rescaling."

In terms of public policy implications, despite recognition of the benefits of cooperation, considerable obstacles mean that economic development efforts to benefit an entire MCR are limited compared to those in smaller metropolitan regions such as Minneapolis-St. Paul and Montpellier. MCRs face many of the obstacles that impede cooperation in smaller metropolitan regions: appointed economic development officials must justify their existence by competing on behalf of their own jurisdiction, rather than pursuing less tangible benefits from metropolitan cooperation; the short time frame of elected officials encourages a preference for visible accomplishments such as groundbreaking and ribbon-cutting in their *own* jurisdiction.

MCRs face additional obstacles to cooperation because of their enormous size. From a spatial perspective, the spillover benefits from a new factory in one locality may benefit most localities within a small metropolitan region. Workers from all parts of a small metropolitan region can easily commute to the facility. In contrast, benefits from the same company's investment may not spillover to an entire MCR because its size prevents easy commuting.

Cooperation to achieve endogenous development, such as infrastructure, is more evident, being relatively straightforward in terms of the costs and benefits. In examples such as the Thames Gateway, cooperation does not necessarily require all MCR localities to participate in order for the entire MCR to benefit.

In contrast, cooperation to attract exogenous development, such as a new company's investment, is more difficult because the costs and benefits are not easy to establish for each jurisdiction. This kind of cooperation does require all MCR localities to participate, raising the question of whether MCRs are just too large and complex for cooperation to attract exogenous development. A successful analogy, however, is the Industrial Development Authority (IDA), which promotes inward investment for Ireland; it controls the process such that localities within its metropolitan regions are left to promote the quality of their infrastructure, labor force, and so on. The IDA exerts authority for the entire country, removing the collective inaction problem that impedes cooperation. Theories of cooperation (for example, iterated game theory) provide insights into the prisoners' dilemma that would otherwise occur. In a simple two-person game, each player chooses to cooperate or defect. The best outcome for both is if they both cooperate; the worst outcome is if they both defect. Yet each opts to defect because neither can guarantee that the other will cooperate.

Rusk's (1995) notion of "elastic" cities may address this prisoners' dilemma; this involves adjusting political boundaries to correspond with the built-up area so that peripheral growth is captured over time. Although Rusk's idea is not realistic for a formal metropolitan tier of government, it may be politically feasible for an economic development agency with responsibility for an entire MCR. This agency would need top-down establishment and bottom-up buy-in, the "teeth" of the IDA, with the pros and cons of elected or appointed members weighed. Although just an idea, any new MCR-wide governance such as this requires more than lip service support so that cooperation moves from paper to practice. While the necessity of providing fewer unnecessary incentives would benefit individual local governments, reducing company relocations that occur purely as a result of public incentives would help the national or supranational economy by minimizing this unproductive use of public and company funds.

A positive aspect of attempting to cooperate like this at the scale of MCRs, ironically, relates to their size—they are "in a league of their own." For smaller metropolitan regions, cooperation may merely shift the scale from a contest between localities to one between metropolitan regions. Unless metropolitan regional competitive advantage is promoted through enhancing conditions for business, and economic specialization occurs between metropolitan regions, competing as a metropolitan region for inward investment eliminates competition only between the jurisdictions *within* particular metropolitan regions—the battle for investment would continue *between* metropolitan regions. Companies may still try to play different metropolitan regions off against each other within the same general area. The smaller number and widely spaced spatial pattern of MCRs, however, and the fact that they compete internationally to a greater extent than smaller metropolitan areas, makes it more difficult for a company to play MCRs off against each other.

Limited MCR cooperation in London, Paris, New York, and Los Angeles, despite the opportunities from already having passed their rapid growth stage and being prosperous relatively in global terms, is partly spatial, due to fragmentation, but is also temporal. These MCRs have existed much longer than those in other parts of the world. Considerable political, economic, and social differences have had more time to develop across these MCRs to undermine cooperation. Consequently, there may be more opportunities for cooperation in parts of the world where the MCRs have developed more recently and where early adoption of cooperation could be instilled, from the top-down and bottom-up, before competition becomes ingrained.

This chapter is not going to conclude with the usual call for more cooperation or an outline of what has already been established is needed for better cooperation—fostering relations of trust, identifying mutual benefits, and so on. Instead, ending on an optimistic note, the recent efforts at improved coordination, albeit top-down by the national governments in the UK, France, and the US, could be viewed as initial steps in a positive trend. Alternatively, finishing on a realistic note, more cooperation is unlikely in the short term unless growing economic, social, and/or environmental concerns take on an urgency that forces localities together (cooperatively) rather than further apart (competitively). Concluding on a pessimistic note, there is no guarantee that such growing concerns will promote more cooperation; they could just as easily result in more competition. Ultimately, the saying that "the whole is greater than the sum of the parts" is useful in thinking about MCR cooperation. For MCRs, however, if the parts cannot come together politically to add up to the whole, then the parts and the whole may be worse off economically, socially, and environmentally.

References

Abu-Lughod, J. (1999) *New York, Chicago, Los Angeles: America's Global Cities*, Minneapolis, MN: University of Minnesota Press.

Brenner, N. (2002) "Decoding the Newest 'Metropolitan Regionalism' in the USA," *Cities*, 19(1): 3–21.

—— (2003) "Metropolitan Institutional Reform and the Rescaling of State Space in Contemporary Western Europe," *European Urban and Regional Studies*, 10(4): 297–324.

Cisneros, H. (1995) *Urban Entrepreneurialism and National Economic Growth*, Washington, DC: US Department of Housing and Urban Development.

Downs, A. (1994) *New Visions for Metropolitan America*, Washington, DC: Brookings Institute.

Fouchier, V. (2008) *Plan Establishment and Application of Decentralized Governance: the Master Plan of the Ile-de-France Metropolitan Region*, Paris: IAURP.

Gainsborough, J. (2003) "Business Organizations as Regional Actors: the Politics of Regional Cooperation in Metropolitan America," *Polity*, 35: 555–72.

Goetz, E. and Kayser, T. (1993) "Competition and Cooperation in Economic Development: a Study of the Twin Cities Metropolitan Area," *Economic Development Quarterly*, 7(1): 63–78.

Hall, P. (2001) "Global City-regions in the Twenty-first Century," in A. Scott (ed.) *Global City-Regions*, Oxford: Oxford University Press, pp. 59–77.

Herrschel, T. and Newman, P. (2002) *Governance of Europe's City Regions*, London: Routledge.

John, P., Tickell, A., and Musson, S. (2005) "Governing the Mega-region: Governance and Networks across London and the South East of England," *New Political Economy*, 10(1): 91–106.

Jonas, A. and Pincetl, S. (2006) "Rescaling Regions in the State," *Political Geography*, 25(5): 482–505.

Jonas, A. and Ward, K. (2007) "New Geographies of Governance, Democracy and Social Reproduction," *International Journal of Urban and Regional Research*, 31(1): 169–78.

Keating, M. (2001) "Governing Cities and Regions: Territorial Restructuring in a Global Age," in A. Scott (ed.) *Global City-Regions*, Oxford: Oxford University Press, pp. 371–90.

Keil, R. (1998) *Los Angeles: Globalization, Urbanization and Social Struggles*, Chichester: Wiley.

Kincaid, J. (1997) *American Cities in the Global Economy: A Survey of Municipalities on Activities and Attitude*, Washington, DC: National League of Cities.

Lefèvre, C. (2003) "Paris—Île-de-France Region," in W. Salet, A. Thornley, and A. Kreukels (eds.) *Metropolitan Governance and Spatial Planning: Comparative Case Studies of European City-Region*, London: Spon, pp. 287–300.

McCarthy, L. (2003) "The Good of the Many Outweighs the Good of the One," *Journal of Planning Education and Research*, 23(2): 140–52.

—— (2004) "The Keep Jeep in Toledo Campaign: a Lost Opportunity for the Wheels of Change?" *Tijdschrift voor Economische en Sociale Geografie*, 95(4): 392–404.

Négrier, E. (2006) "Rescaling French Urban Territories," *European Planning Studies*, 14(7): 939–58.

New York Times (1996) "The Wrong Way to Lure Business," *New York Times Editorial*, September, Section 4: 16.

Ohmae, K. (1995) *The End of the Nation State*, New York: The Free Press.

Pastor, M., Dreier, P., Grigsby, J., and López-Garza, M. (2000) *Regions that Work: How Cities and Suburbs Can Grow Together*, Minneapolis, MN: University of Minnesota Press.

Ross, B. and Levine, M. (2001) *Urban Politics*, 6th edition, Itasca, IL: F. E. Peacock.

Rusk, D. (1995) *Cities Without Suburbs*, 2nd edition, Washington, DC: Woodrow Wilson Center.

Salet, W., Thornley, A., and Kreukels, A. (2003) *Metropolitan Governance and Spatial Planning: Comparative Case Studies of European City-regions*, London: Spon.

Sassen, S. (2000) *Cities in a World Economy*, 2nd edition, Thousand Oaks, CA: Pine Forge.

Scott, A. (ed.) (2001) *Global City-Regions*, Oxford: Oxford University Press.

Stecker, S. and Steinberg, D. (2007) *Pay, Or We (Might) Go: How Citigroup Games the States and Cities*, Trenton, NJ and New York, NY: NJ Policy Perspective/Good Jobs NY.

Subra, P. and Newman, P. (2008) "Governing Paris—Planning and Political Conflict in Ile-de-France," *European Planning Studies*, 16(4): 521–35.

Walker, D. (1987) "Snow White and the 17 Dwarfs," *National Civic Review*, 76(1): 14–28.

8 Regions of cities

Metropolitan governance and planning in Australia

John Abbott

Introduction

Australia is a highly urbanized country, with 89 percent of the population living in urban centers of greater than 1,000 persons (UN Habitat 2009: 244). Australia's three largest metropolitan regions—Sydney, Melbourne, and Brisbane (South East Queensland)—are not large by international standards (UN Habitat 2008: 238–47). However, because of their low residential densities, they cover large areas, include multiple centers, and are developing as regions of cities.

O'Connor (1993) identified three potential mega-cities in Australia in 1993: Sydney, Newcastle, and Wollongong; Melbourne and Geelong; and Brisbane, the Sunshine Coast, and the Gold Coast. This chapter will explore the different contexts, circumstances, and patterns of growth of these three metropolitan regions and how they relate to the concept of a mega-city region as a cluster of contiguous, administratively separate, and highly networked centers or cities, as discussed in Chapter 1 of this book and in Hall and Pain (2006). It will explore how the patterns of growth in the three metropolitan regions are a result of historical growth factors, governance arrangements, and planning intentions. The rationales of governance and planning, along with the distribution of power among stakeholders, will be examined. These governance and planning arrangements, and complexity and uncertainties about metropolitan futures, will affect how these regions develop in the future.

Government, governance, growth, and metropolitan planning

Australia's metropolitan regions have developed within a federal system of government with three levels: the Commonwealth; the states; and local government. The national government, or Commonwealth of Australia, sets the context for metropolitan growth with control of immigration, taxation, and the national economy, along with funding of national infrastructure projects. The state governments are the "dominant force" and major players in metropolitan regions, and provide transport, health, education, infrastructure, and services, and are generally responsible for metropolitan planning strategies (Forster 2004: 141). Local governments provide local roads, water, and cultural services, and prepare local land use plans. They are important players in planning for particular cities within metropolitan regions. However, as Forster (2004: 143) says, "the most significant aspect of local government is its fragmented nature." All three levels of government are involved in the government and planning of Australia's metropolitan regions.

If the concept of government is widened to that of governance, then the roles of the community and of the private sector, or market, are also important in the ongoing management and planning of metropolitan regions (Phares 2004). Gleeson and Low (2000: 137) note that community participation is now mandated and widespread in urban and metropolitan planning in Australia. In many instances, it can be tokenistic, but it sometimes involves "genuine [. . .] empowerment" of the community, such as in the Western Suburbs Action Plan in Melbourne. The private sector has always sought to influence governments in order to get approval for their projects and to gain financial support from the government. With the rise in influence of neoliberal ideas from the 1980s onwards, many people wanted to replace government services with private-sector provision. Gleeson and Low (2000: 182) have expressed concerns that under some conservative (Coalition) governments in Australia, private-sector development interests have been setting the agenda for government discussions on property and planning issues.

Table 8.1 shows the population and growth rates of Australia's metropolitan regions, which are based on the latest census information (2006) and O'Connor's boundaries (1993). South East Queensland (SEQ) was the fastest growing region over the period 1996 to 2006, in percentage and in absolute terms, and this rapid growth is continuing. Of the three largest metropolitan regions, Melbourne was the second fastest growing region, in percentage terms, and Sydney was the slowest.

Population projections for Australia and for the metropolitan regions to 2031, prepared by the Australian Bureau of Statistics (ABS), are shown in Table 8.2. These projections cover only the main statistical division (SD) in each metropolitan region. The projections were updated in September 2008 to reflect the Commonwealth Government's policy of higher levels of international migration and higher fertility rates in Australia. Over this period, the fastest growing metropolitan region (in percentage terms) is expected to be Perth, with SEQ second. However, these figures do not include the fast-growing SEQ SDs of the Gold Coast and the Sunshine Coast.

The growth and patterns of development of all metropolitan regions in Australia are guided and controlled by metropolitan plans and strategies, generally to 2031, as indicated in Table 8.3. In all cases except for Hobart, these have been prepared by the planning department of the relevant state government. Local government involvement in the preparation of metropolitan plans varies from informal consultation to formal processes of partnership, as in the case of South East Queensland.

Table 8.1 Greater metropolitan regions in Australia: population growth 1996 to 2006

Metropolitan region (statistical divisions)	*Population 1996*	*Population 2006*	*Absolute growth 1996–2006*	*Growth 1996–2006 (%)*
Sydney (Sydney, Hunter, Illawarra)	4,808,237	5,316,379	508,142	10.57
Melbourne (Melbourne, Barwon)	3,522,797	4,014,361	491,564	13.95
South East Queensland (Brisbane, Gold Coast, Sunshine Coast, West Moreton)	2,138,215	2,706,302	568,087	26.57
Perth (Perth)	1,295,092	1,519,510	224,418	17.33
Adelaide (Adelaide, Outer Adelaide)	1,182,768	1,275,041	92,273	7.80
Hobart (Greater Hobart)	195,718	205,566	9,846	5.03

Source: ABS 2007.

Table 8.2 Metropolitan regions in Australia: population projections to 2031

Metropolitan region (statistical divisions)	*Population 2006*	*Population 2031*	*Absolute growth 2006–2031*	*Growth 2006–2031 (%)*
Sydney (Sydney)	4,282,000	5,710,900	1,428,900	33.37
Melbourne (Melbourne)	3,743,000	5,355,200	1,612,200	43.07
South East Queensland (Brisbane)	1,819,800	2,902,000	1,082,200	59.47
Perth (Perth)	1,518,700	2,456,700	938,000	61.76
Adelaide (Adelaide)	1,145,800	1,438,800	293,000	25.57
Hobart (Greater Hobart)	205,500	253,200	47,700	23.21
Australia	**20,697,900**	**28,786,500**	**8,088,600**	**39.08**

Source: ABS 2008.

Table 8.3 Metropolitan plans and strategies in Australia

Metropolitan region (state)	*Name of current metropolitan plan (date of release)*	*Prepared by*
Sydney (New South Wales)	City of Cities: A Plan for Sydney's Future (2005)	Department of Planning, New South Wales Government
Melbourne (Victoria)	Melbourne 2030: Planning for Sustainable Growth (2002)	Department of Infrastructure, Victorian Government
South East Queensland (Queensland)	South East Queensland Regional Plan 2009–2031 (2009)	Department of Infrastructure and Planning, Queensland Government
Adelaide (South Australia)	Planning Strategy for Metropolitan Adelaide (2007)	Department of Planning and Local Government, Government of South Australia
Perth (Western Australia)	Network City: Community Planning Strategy for Perth and Peel (2004)	Department for Planning and Infrastructure, Government of Western Australia
Hobart (Tasmania)	Hobart 2025: A Strategic Framework (2007)	Hobart City Council

Source: Metropolitan plans listed above.

Issues that these metropolitan plans seek to influence and guide include: managing the impact of rapid population and economic growth; enhancing global economic competitiveness; achieving a more compact pattern of growth focusing on infill and development around major centers; integrating transport and land use planning based on transit-oriented development principles; promoting employment growth in regional cities and major centers; matching the sequence of growth and the provision of infrastructure; enhancing green space; protecting good agricultural land; addressing climate change; improving governance; and promoting sustainability.

The following sections will explore the patterns of growth in Australia's three largest metropolitan regions and link these to historical factors, governance arrangements, and planning intentions.

Governance and planning in Sydney

Sydney is Australia's largest metropolitan region, as indicated in Table 8.1. It is also Australia's first city in a number of other ways: the Sydney Cove penal colony was the first urban settlement established in Australia in 1788; Sydney was the first region to produce an approved metropolitan scale urban plan; and it is Australia's first global city (Searle 1996; Newman and Thornley 2005).

The first Sydney Municipal Council was established in 1842. By 1862, ten other small municipalities had been established, covering the rapidly expanding Sydney urban area, with a total population of around 100,000 people (Spearritt and DeMarco 1988: 3). This set the pattern of an urban area consisting of numerous small local governments that continues until the present day. At the end of the Second World War in 1945, a second tier of local government was established in Sydney, called the Cumberland County Council (CCC). It consisted of ten members elected by the then sixty-nine constituent councils of the Sydney region, and its main purpose was to prepare a planning scheme for the Sydney area (Howe 2000). The *Planning Scheme for the County of Cumberland, New South Wales* (the County of Cumberland Plan) (CCC 1948) was released in 1948 and approved in 1951. It was modeled on the Greater London Plan of 1944, and it included a green belt to constrain the westward growth of the city, but did not include any provisions for satellite cities (Howe 2000). The County of Cumberland Plan severely underestimated the scale of postwar immigration to Australia and the rapid growth in the population of Sydney. The Council "resisted Ministerial pressure to release more and more of the green belt for housing" (Spearritt and DeMarco 1988: 25). Because of this ongoing conflict, the New South Wales (NSW) government abolished the Cumberland County Council in 1964 and took control of metropolitan planning for Sydney. Since then, there have been no further attempts to introduce metropolitan-wide regional government, and the state government has retained direct control of Sydney's growth through its planning agencies and a series of metropolitan plans.

Work on a new plan to replace the County of Cumberland Plan commenced in 1964, under the direction of the newly established State Planning Authority (SPA). The NSW Government released the *Sydney Region Outline Plan 1970–2000 AD* in 1968 (the 1968 Sydney Plan) (SPA 1968). The 1968 Sydney Plan identified four growth corridors radiating out from the Sydney central business district (CBD): a western corridor commencing at the sub-regional center of Parramatta; a southwestern corridor centered on a second sub-regional center at Campbelltown; and two less clearly defined corridors to the north and northwest. Economic and employment growth would be encouraged at the two sub-regional centers to act as "counter-magnets to the CBD" (Spearritt and DeMarco 1998: 31). By the middle of the 1980s, growth was continuing, and land identified for urban growth in the 1968 Sydney Plan was rapidly being exhausted. A new metropolitan planning process commenced and *Sydney Into Its Third Century* (the 1988 Sydney Plan) was released in 1988 (DEP 1988). Because of the shortage of land and the costs of providing infrastructure on the urban fringe, the 1988 Sydney Plan focused on a strategy of "concentrated" growth (Lennon 2000: 154). It identified three regional centers—Sydney CBD, North Sydney, and Parramatta—and noted that the NSW government would "continue to promote Parramatta as the major center for the relocation of government offices" (DEP 1988: 51).

The election of a state coalition government in 1988 saw the commencement of a new metropolitan planning process and the release of the strategy called *Cities For the 21st Century* (the 1995 Sydney Plan) in 1995 (DOP 1995). The 1995 Sydney Plan claimed to be a "new approach" with an "emphasis on integrated, whole of government urban management" (DOP 1995: 11). The Sydney metropolitan region was defined widely in this plan and included the adjacent regional cities of Newcastle (170 kilometers north of the Sydney CBD) and Wollongong (85 kilometers south of Sydney). The reasons for their inclusion were not clearly explained. The 1995 Sydney Plan identified four primary centers—Sydney CBD, Parramatta, Newcastle, and Wollongong. It explicitly recognized the local planning role of the fifty-three local governments in the metropolitan region and the impact of Commonwealth Government policies about the national economy, transport, and immigration on the region (DOP 1995: 5). Searle (2004: 373) noted that the 1995 Sydney Plan was based on a "neoliberal agenda" and accorded a greater role and "priority to private investors and the market generally." Another change, back to a state Labor government in 1995, saw the initiation of a metropolitan planning process so that the new government could have "its own strategy" (Searle 2004: 376). The new metropolitan strategy, released in 1998, was called *Shaping Our Cities: The Planning Strategy for the Greater Metropolitan Region of Sydney, Newcastle, Wollongong and the Central Coast* (the 1998 Sydney Plan) (DUAP 1998). In relation to cities and centers, this plan was similar to its 1995 predecessor. It was based on the wider metropolitan region, including Newcastle and Wollongong; and again identified four primary centers as Sydney

CBD, Parramatta, Newcastle, and Wollongong. It also continued the neoliberal agenda of working closely with the private sector (Searle 2004).

The current plan for the Sydney metropolitan area is the *City of Cities: A Plan for Sydney's Future* (the City of Cities Plan) (DOP 2005). Although no reason was given, the City of Cities Plan reverts to a more traditional census-based geography of the Sydney metropolitan region, and does not include Newcastle and Wollongong. The metropolitan region is similar to the Sydney Statistical Division; however, it does not include the Central Coast area, centered on Gosford. The City of Cities Plan explicitly identifies a vision for Sydney at 2031 and sets out proposals for "stronger cities within the metropolitan area" (DOP 2005: 8). It identifies five cities: Sydney City (the historical CBD) and North Sydney support the role of Sydney as a global city through internationally focused business, tourism, culture, research and education, and entertainment activities and facilities; Parramatta, Liverpool, and Penrith provide for more locally focused business, shopping, health, education, and cultural activities and facilities. These three western cities provide for service and manufacturing sector jobs closer to the growing residential areas of Sydney. The City of Cities Plan implicitly identifies Gosford as the primary center for the Central Coast area and as a sixth city within the Sydney SD. While Sydney City, North Sydney, and Parramatta have been identified in a number of previous plans, little evidence or explanation is provided in the City of Cities Plan for the inclusion of Liverpool and Penrith as regional cities (Bunker and Holloway 2006: 6).

Implementation and governance arrangements for the City of Cities Plan, as for all Sydney metropolitan plans since 1968, continue to rely on a central and powerful role for the NSW government, through the Department of Planning (DOP) and other State agencies. Key implementation activities for the City of Cities Plan include (DOP 2005: 252):

- preparation of a state infrastructure strategy to guide state infrastructure investment priorities for the Sydney region;
- state involvement in "Strategic Places," including Growth Center release areas (through the Growth Centers Commission) and urban renewal sites (through Landcom); and
- sub-regional and local planning.

Ten sub-regional areas have been identified as a basis for sub-regional planning to be led and coordinated by the DOP. The aim of this process is to translate the metropolitan strategy into an intermediate level sub-regional strategy for groups of local governments. A working group of local governments, state agencies, communities, and private-sector stakeholders was established to advise the minister on the overall sub-regional planning process. Local governments are to be involved in the relevant sub-regional planning process for their area. However, the City of Cities Plan does not clarify their role and powers, the status of their local land use, strategic or center plans, and ways of resolving any conflicting intents. Similarly, any direct role for the private sector and land development companies in this process is not clear.

As discussed, Sydney has an existing and evolving role as a leading global city. Newman and Thornley (2005: 187) note that Sydney has "established itself as the leading Australian city in world terms" through its role as the major international air travel hub (and base of Qantas), as the most important financial center in Australia, and as the "location for many transnational corporations wanting to serve South East Asia." The City of Cities Plan identifies the cities of North Sydney and Sydney City and the crescent-shaped corridor through these from Chatswood to the Sydney CBD and on to the Sydney Airport and Port Botany as Sydney's "global economic corridor" (Newman and Thornley 2005: 39). It identifies and facilitates development in this corridor and does not seek to change the trend.

The governance of the Sydney metropolitan region is characterized by: strong overall state government control; the development of state growth management institutions, such as the Growth Centers Commission; a lack of direct involvement by the Commonwealth Government; and an active but circumscribed involvement by local government. Planning for the Sydney region has long recognized the need to support a network of centers and cities, but the location and number of these has varied in different plans. The boundaries of the metropolitan region, and the inclusion of the adjacent regional cities of Newcastle and Wollongong, have also varied in different plans. As links to these adjacent cities grow, Sydney is developing into a mega-city region. This is occurring in the context of strong state government control, and the planning focus can be readily adapted to the larger region in the future.

Governance and planning in Melbourne

Melbourne is the second largest metropolitan region in Australia and has vied with Sydney for growth, influence, and prestige for many years. The settlement of Port Phillip was established as part of the colony of New South Wales. It was surveyed and laid out on a "colonial urban grid" and renamed Melbourne in 1837 (Proudfoot 2000: 12–17). The discovery of gold at Ballarat and Bendigo in 1851 caused the population of Melbourne to explode. By 1901, it had a population of around 500,000 people and was larger than Sydney (Proudfoot 2000). The Melbourne International Exhibition of 1880 raised the status of Melbourne as an international city and—along with its lavish Victorian buildings—gained it the title of "Marvelous Melbourne" (Davison 1978: 229).

Like Sydney, the growing urban area of Melbourne comprised numerous small, local governments. The release of the County of Cumberland Plan for Sydney in 1948 "highlighted Melbourne's failure to make progress on a metropolitan plan" or on metropolitan government (Howe 2000: 92). In 1949, the Victorian government passed legislation to give the metropolitan water and sewerage authority—a state authority—the responsibility to prepare a metropolitan plan. The Melbourne and Metropolitan Board of Works (MMBW) included fifty-three members representing metropolitan local governments (Howe 2000: 95). In effect, the MMBW was a second-tier metropolitan level of government responsible for water and sewerage infrastructure, and metropolitan planning.

The MMBW team of engineers and planners worked quickly to produce the *Melbourne Metropolitan Planning Scheme* (the 1954 Melbourne Plan) in 1954 (MMBW 1954). McLoughlin (1992: 40) notes that the production of a "relatively sophisticated plan in only three years is remarkable." The 1954 Melbourne Plan was a zoning plan and associated regulatory ordinance that controlled land use and development in its own right and provided a regulatory context for local government planning schemes. It identified the Melbourne CBD, and five second-tier "District Business" centers at Box Hill, Dandenong, Footscray, Moorabbin, and Preston. It limited urban sprawl by surrounding the built-up area with a rural zone, which had a relatively high minimum lot size for subdivision. Importantly, it identified "public purpose reservations" for future roads, open space wedges, and linear parks along rivers and creeks (McLoughlin 1992: 43–44). There were numerous public objections to the 1954 Melbourne Plan that had to be worked through, and it was not formally approved by the state government until 1968.

Like the County of Cumberland Plan, the 1954 Melbourne Plan significantly underestimated the rate of postwar immigration, and thus the population growth in Melbourne. Redevelopment of inner city areas was not occurring, and there was considerable pressure to open up more land for housing development on the urban fringe. Before the 1954 Metropolitan Plan was approved, the MMBW had begun work on a review (Morison 2000). It produced its "Mark Two corridors and wedges plan" for the metropolitan area (Huxley 2000: 141), called *Planning Policies for the Melbourne Metropolitan Region* (the 1971 Melbourne Plan), in 1971 (MMBW 1971). The policy intent of the 1971 Melbourne Plan was to direct outward metropolitan expansion to the north and west, rather than the south and east, and for this growth to be in the form of urban corridors, open space wedges, and satellite cities. It also proposed a "massive new freeway grid," which resulted in "intense public opposition" and the eventual removal of major sections of the proposed freeways (McLoughlin 1992: 52). The 1971 Melbourne Plan supported the Melbourne CBD but did not identify any other centers or include any centers policies (McNabb 2001: 32). After consideration of objections and advice from other state agencies and boards, the government approved the 1971 Melbourne Plan in 1974. Further policy reviews by the MMBW in the late 1970s resulted in the release of a "Mark III version" of the metropolitan plan and policies in 1980 (McLoughlin 1992: 61). This was called the *Metropolitan Strategy* (the 1980 Melbourne Plan) (MMBW 1980). It was based on a policy of "incremental growth" and identified about twenty second-order "activity centers." McLoughlin (1992: 61–62) notes that the 1980 Melbourne Plan was weak in relation to how the activity centers policy would be implemented.

The governance arrangements for metropolitan planning in Melbourne in the 1970s and early 1980s were complex, with the state government receiving planning advice from the MMBW, in addition to the Town and Country Planning Board (TCPB). The TCPB was a statutory board established in 1944 to provide advice on planning scheme amendments, including amendments to the Melbourne Metropolitan Planning Scheme. Frequently the advice of these two bodies was in conflict, and may also have clashed with the views of local governments, such as the Melbourne City Council (McLoughlin 1992: 53, 64). In 1985 the state government moved to resolve this situation by transferring the planning powers and planners from the MMBW and the TCPB to a new planning agency, called the Ministry for Planning and Environment (MPE) (McLoughlin 1992: 76). The MPE commenced work on a new metropolitan plan and *Shaping Melbourne's Future* (the 1987 Melbourne Plan) was released in 1987 (MPE 1987). The 1987 Melbourne Plan maintained the urban growth corridors and had a major focus on urban consolidation. It supported the Melbourne CBD and identified seven "strategic district centers" in the Melbourne metropolitan region, at Frankston, Dandenong, Ringwood, Greensborough, Broadmeadows, Sunshine, and Werribee. An eighth strategic district center was identified at Geelong, outside the formal metropolitan region (Lennon 2000: 157–59).

With the election of a new Liberal state government in 1992, Melbourne's metropolitan planning strategy was separated into two areas: the central city; and suburban neighborhoods and the rest of the region. Proposals for the central city, including the replacement of the Melbourne City Council with state-appointed commissioners, focused on promoting economic development through private-sector waterfront projects. Proposals for the region were expressed in *Living Suburbs: A Policy for Metropolitan Melbourne into the 21st Century* (DPD 1995). It focused on "improving Melbourne's business environment and competitiveness" and supported a network of activity centers located at transport nodes (Hamnett 2000: 178).

The return of a state Labor government in 1999 led to the initiation of a new, integrated metropolitan planning process and the release of the current metropolitan plan for Melbourne, called *Melbourne 2030: Planning for Sustainable Growth* (the Melbourne 2030 Plan) in 2002 (DOI 2002). The Melbourne 2030 Plan is a very wide-ranging document that focuses on sustainability. It identifies nine strategic directions, as follows:

- a more compact city;
- better management of metropolitan growth;
- networks with the regional cities;
- a more prosperous city;
- a great place to be;

- a fairer city;
- a greener city;
- better transport links; and
- better planning decisions, careful management.

In seeking to achieve a more compact city, the Melbourne 2030 Plan identifies the Melbourne CBD as the central activity district, nine transit cities, and twenty-five principal activity centers as the focus for high-quality and higher-density development. It establishes an urban growth boundary to contain and direct Melbourne's outward expansion, and continues the policy of protecting green wedges.

Implementation and governance arrangements for the Melbourne 2030 Plan are based on a strong and direct role for the state government through the Department of Infrastructure (DOI) and its successors. The plan talks about the critical need to "establish and maintain an ongoing partnership" with local government (DOI 2002: 152). It provides little evidence that this occurred during its preparation, and lacks proposals for how it will occur during implementation. However, the implementation of key policies, such as activity centers, the urban growth boundary, green wedges, and growth areas rely heavily on the negotiated or forced agreement of local governments to amend their planning schemes. Critics of the Melbourne 2030 Plan, and of the activity centers policy in particular, note that "much is expected [. . .] although the sole implementation mechanisms are local scale statutory plans." Councils complained about their lack of involvement in developing the policy and the imposition of "an activity center template across a large number of very different actual and potential centers" (Birrell et al. 2005: 02/2, 02/25).

Melbourne and Geelong (a regional city, 70 kilometers southwest of Melbourne around Port Phillip Bay, with a population of 270,000 in 2006) have always been governed and planned separately. However, the Melbourne Plan makes a number of proposals that indicate that links between the two urban areas are strengthening or need to be strengthened. These include proposals for: a fast passenger rail link; improved freight rail access to the Port of Geelong; an urban growth boundary around Geelong; and protection of Avalon airport, regional parks, and green recycling areas in the inter-urban break between Melbourne and Geelong.

One of the strategic directions in the Melbourne 2030 Plan is "a more prosperous city," and this involves "creating opportunities for innovation and the knowledge economy" (DOI 2002: 73). The spatial aspects of this policy are less well developed than in Sydney, although the importance of the Central Melbourne CBD is emphasized. Yigitcanlar et al. (2008: 70) have noted the importance of Melbourne as one of the "emerging knowledge cities around the world." However, they also noted that cultural factors and the presence of research institutions are more important than spatial policies in supporting knowledge cities.

The governance of the Melbourne metropolitan region is characterized by: strong overall state government control; a lack of involvement by the Commonwealth Government; and an active, yet limited involvement by local government. The Melbourne 2030 Plan makes few proposals for new institutional arrangements for growth management. Its proposals for a network of transit cities and activity centers have been criticized for proposing too many centers for these to be effective (Birrell et al. 2005). Links with the adjacent regional city of Geelong are increasing and Melbourne is developing into a mega-city region covering both Melbourne and Geelong. Given the strong state government control, it will not be difficult at some stage in the future to shift the focus of planning to a metropolitan region covering both cities.

Governance and planning in South East Queensland

South East Queensland (SEQ) is the third largest metropolitan region in Australia, and is centered on the City of Brisbane. Brisbane City Council is the largest and the most powerful local government in Australia, and rivals the state government for influence in SEQ planning.

Brisbane was established as another penal settlement within the colony of New South Wales in 1824. The colony grew rapidly, and the Municipal Council of Brisbane was established in 1859 (Greenwood and Laverty 1959). As in Sydney and Melbourne, the expanding urban area of Brisbane soon consisted of a large number of small local governments. However, in Brisbane in the early 1900s, support grew for "the creation of a greater city" and the advantages of this were seen to be "efficiency and economy in administration, uniformity of approach [. . .] and an improved system of representation" (Greenwood and Laverty 1959: 450). The City of Brisbane Act 1924 was passed in late 1924 and the Greater Brisbane Council (GBC) was established with wide-ranging powers in early 1925. It included all or parts of twenty former local governments. Following the passage of town planning legislation by the state in 1934, the GBC appointed Ronald McInnes to prepare a town plan for the city and this was completed by 1944. However, inter-government rivalry intervened and the plan was "not given statutory authority because of the Queensland Government's determination to restrict the power of the GBC" (Howe 2000: 84). The first town plan for Brisbane did not come into effect until 1965.

In the 1970s, following initiatives from the Commonwealth Government, the Queensland government established ten regional coordination councils (RCCs) to do regional planning. These included the Moreton Regional Coordination Council (Moreton RCC), which covered Brisbane City and sixteen surrounding municipalities. One of the main projects the Moreton RCC was involved in was the preparation of a Moreton Region Growth Strategy, which commenced in 1974. The strategy preparation process was completed in eighteen months, and it was released as a ten-volume report entitled the *Moreton Region Growth Strategy* (the Moreton Region Strategy) in June 1976

(COG and Cities Commission 1976). The Moreton Region Strategy was a comprehensive metropolitan strategy for SEQ, covering land use, infrastructure, and economic and social issues. It proposed redirecting population growth from Brisbane City to (what is now) Logan, the Gold Coast, and the Sunshine Coast. A "fundamental proposition" of the strategy was that "the City of Brisbane is now at a size where development of other centers offering alternative employment locations to the central area is both necessary and feasible" (COG and Cities Commission 1976: 68). It proposed developing subregional centers at Ipswich, the Gold Coast, and Logan. There was no agreed process of implementation of the Moreton Region Strategy and there was a lack of political support from local governments, the state government and the new Commonwealth Government. In 1977, after the 1975 removal of the Whitlam Commonwealth Government, the state government abolished all the RCCs, including the Moreton RCC (Harris 1978: 192). However, the seventeen local governments that had been involved agreed to continue to meet under their own chair, and formed the Moreton Regional Organization (MRO) of Councils in 1977. The perceived failure of regional planning and coordination in the 1970s influenced state and local governments' negative attitudes to regional planning through the 1980s (Abbott 2001).

During the 1970s and 1980s, Queensland had a conservative coalition state government that saw planning as a local government matter but reserved the right to intervene by direct rezonings to progress its own development projects or those of its private-sector partners. Findings of widespread corruption in the state government in the late 1980s led to the election of a new state Labor government at the end of 1989.

The new state government saw the need for a regional planning approach to address the impact and problems of rapid population growth in SEQ. They agreed to work in a "voluntary partnership" approach—called SEQ 2001 —with local governments and the Commonwealth Government to prepare an SEQ regional planning strategy (Abbott 2001: 114; also Abbott 1995). The *South East Queensland Regional Framework for Growth Management 1995* (the RFGM 1995 Plan) was endorsed by all three levels of government (Commonwealth, state, and local government) in 1995 (RCC 1995). The RFGM 1995 Plan was a broad, non-statutory policy plan that covered land use, environmental, and social and economic aspects of managing growth. It recognized that SEQ included "four distinctive urban areas—the Brisbane Metropolitan Area, the Sunshine Coast, the Gold Coast, and Toowoomba" (RCC 1995: 14), and identified seven key centers within these urban areas. Preparation of the RFGM 1995 Plan also resulted in the development of ongoing institutional arrangements for metropolitan planning in SEQ, reflecting the partnership approach between the state and local governments. The most important formal aspect of this partnership was the establishment of an SEQ Regional Coordination Committee (RCC), consisting of a number of ministers and mayors, as a high-level advisory committee to governments (Abbott 2001: 18–19).

Weaknesses in the implementation of the RFGM 1995 Plan, and subsequent non-statutory plans in 1998 and 2000, in the context of ongoing rapid regional growth, led to agreement by both state and local governments that a statutory regional plan was needed in SEQ. The state government established a powerful new agency, the Office of Urban Management (OUM), to prepare the plan in partnership with local governments. The first statutory metropolitan plan for SEQ, called the *South East Queensland Regional Plan 2005–2026* (the 2005 SEQ Plan) was released in 2005 (OUM 2005a). The 2005 SEQ Plan recognized the four urban areas in SEQ and defined their agreed extent in a statutory "Urban Footprint" (OUM 2005a: 16). It also identified the Brisbane CBD as the primary activity center for the region and fifteen principal activity centers.

The 2005 SEQ Plan was to be implemented in partnership with local governments through the RCC. The plan acted directly to prohibit urban development outside the Urban Footprint, and supported compact infill development around activity centers within the Urban Footprint. Local governments were to prepare local growth management strategies (LGMSs) showing how the regional policies and dwelling targets were to be met in their areas. The state government was to prepare an annual regional infrastructure program linked to the state budget and reflecting the policies and priorities of the 2005 SEQ Plan. The first *South East Queensland Infrastructure Plan and Program 2005* (SEQIPP 2005) was released in 2005 (OUM 2005b). The 2005 SEQ Plan had proposed that it should be formally reviewed every five years. The state government decided to bring this forward by a year and the current *South East Queensland Regional Plan 2009–2031* (the 2009 SEQ Plan) was released in 2009 (DIP 2009). The 2009 SEQ Plan continues the broad planning approach of the 2005 SEQ Plan, but gives more emphasis to emerging issues such as sustainability and climate change. The requirement to prepare LGMSs has been removed, and has been replaced by more detailed "sub-regional narratives" (DIP 2009: 17).

One of the "desired regional outcomes" of the 2009 SEQ Plan is a "strong, resilient and diversified economy," and the plan notes the importance of "creating high skilled jobs in knowledge-based industries" (DIP 2009: 111). It identifies science and technology employment opportunity areas that need to be protected from incompatible land uses. A related report, entitled *Smart Cities: Rethinking the City Centre*, identifies a "knowledge corridor" linking universities, hospitals, and other research institutes with the University of Queensland (St. Lucia), the Brisbane CBD, and the Royal Brisbane Hospital (SSC 2007). These sites are linked by a busway on an exclusive right of way.

The governance of the SEQ metropolitan region is characterized by: strong overall state government control; an active planning partnership with local governments through the RCC; and a lack of involvement by the Commonwealth Government (although it is represented on the RCC). The 2009 SEQ Plan recognizes the four separate urban areas within SEQ, and the strong and increasing links between them. Although SEQ is smaller in population than Sydney or Melbourne, it can already be characterized as a mega-city region; governance and planning arrangements are in place covering the whole mega-city region.

Complexity and uncertainties in metropolitan governance and planning

Metropolitan growth and planning occurs as part of a "complex adaptive system" with "multiple organizations, plans, issues and actions interacting in complex ways" (Abbott 2009: 503). Salet and Thornley (2007: 188) note that metropolitan planning strategies "face extremely dynamic conditions of complexity and variety." However, in conditions of complexity and uncertainties, good governance and planning become even more important (Newman and Thornley 2005: 276; Ross and Doyle 2009).

The governance and institutional arrangements of metropolitan regions and planning outcomes are interrelated, and the metropolitan plans produced are "in many ways, inherent in the organizational and political context that produced them" (Abbott 2009: 515). Alexander (2009: 518) says that many existing institutions are "survivors from a simpler past" and that planning in complex multi-organizational systems, such as metropolitan regions, "usually involves some institutional transformation or change to enable action in this intrinsically changing and uncertain context."

This review of governance and planning arrangements in Australia's three largest cities shows that, in all cases, the relevant state government is strongly in control of the process. But governments (in this case state governments) cannot "simply decide" on the growth of metropolitan regions; because of change, complexity, and uncertainties, they require the involvement of other stakeholders in the planning process to provide relevant technical, political, organizational, and economic information (Salet and Thornley 2007: 192). In Sydney, new institutional arrangements were developed in the form of the Growth Centers Commission to involve relevant stakeholders in the planning of local growth areas. In SEQ, the RCC provides opportunities for local government involvement and input. In Melbourne, the development of new institutional arrangements to support the Melbourne 2030 Plan seems to be lacking. Bunker and Searle (2007: 619) feel that the current Sydney and Melbourne plans are "too prescriptive and deterministic in their provisions but they do provide a basis for adaptation and adjustment to changing conditions." In all three regions, linkages to the Commonwealth Government need to be improved. Linkages to the private sector and to the community on a regional basis, rather than a project or local basis, also need to be enhanced.

Governance arrangements and planning processes need to be considered together. New governance arrangements may need to be established before metropolitan planning can commence or may need to be developed as part of the planning process and as a necessary part of implementation of the plan.

Salet and Thornley (2007: 191) feel that this is not a matter of establishing "encompassing territory government" but rather one of "organizing connectivity" between key stakeholders and planning activities. Planning processes for metropolitan regions and developing mega-cities in Australia have the capacity to do this, but state governments need to be more open, innovative, and flexible in involving other important stakeholders.

Acknowledgment

The assistance of Dr. Alison Taylor, Queensland Treasury, with the population figures and projections in this chapter is greatly appreciated.

References

Abbott, J. (1995) "SEQ 2001: Quality Strategic Planning for South East Queensland," *Australian Planner*, 32(3): 135–38.

—— (2001) "A Partnership Approach to Regional Planning in South East Queensland: Ten Years of SEQ 2001," *Australian Planner*, 38(3/4): 114–20.

—— (2009) "Planning for Complex Metropolitan Regions: A Better Future or a More Certain One?" *Journal of Planning Education and Research*, 28(4): 503–17.

ABS (Australian Bureau of Statistics) (2007) *Regional Population Growth Australia 1996–2006*, Catalogue No. 3218, Canberra: Australian Bureau of Statistics.

—— (2008) *Population Projections Australia 2004–2101*, Catalogue No. 3222, Canberra: Australian Bureau of Statistics.

Alexander, E. (2009) "Planning in Complexity: Institutional Design Implications," *Journal of Planning Education and Research*, 28(4): 518–24.

Birrell, B., O'Connor, K., Rapson, V., and Healy, E. (2005) "Concentrating Melbourne: the Activity Centre Strategy," in B. Birrell, K. O'Connor, V. Rapson, and E. Healy (eds.) *Melbourne 2030: Planning Rhetoric Versus Urban Reality*, Melbourne: Monash University ePress, pp. 02/1–02/28.

Bunker, R. and Holloway, D. (2006) *How Much of "City of Cities, A Plan for Sydney's Future" is Likely to Happen?* City Futures Research Centre, Issues Paper No. 2, Sydney: University of New South Wales.

Bunker, R. and Searle, G. (2007) "Seeking Certainty: Recent Planning for Sydney and Melbourne," *Town Planning Review*, 78(5): 619–42.

CCC (Cumberland County Council) (1948) *Planning Scheme for the County of Cumberland, New South Wales*, Sydney: Cumberland County Council.

COG and Cities Commission (1976) *Moreton Region Growth Strategy*, Brisbane: Coordinator General's Department, Queensland Government.

Davison, G. (1978) *The Rise and Fall of Marvelous Melbourne*, Melbourne: Melbourne University Press.

DEP (Department of Environment and Planning) (1988) *Sydney into Its Third Century: Metropolitan Strategy for the Sydney Region*, Sydney: Department of Environment and Planning, NSW Government.

DIP (Department of Infrastructure and Planning) (2009) *South East Queensland Regional Plan 2009–2031*, Brisbane: Department of Infrastructure and Planning, Queensland Government.

DOI (Department of Infrastructure) (2002) *Melbourne 2030: Planning for Sustainable Growth*, Melbourne: Department of Infrastructure, Victorian Government.

DOP (Department of Planning) (1995) *Cities for the 21st Century: Integrated Urban Management for Sydney, Newcastle, the Central Coast and Wollongong*, Sydney: Department of Planning, NSW Government.

—— (2005) *City of Cities: A Plan for Sydney's Future*, Sydney: Department of Planning, NSW Government.

DPD (Department of Planning and Development) (1995) *Living Suburbs: A Policy for Metropolitan Melbourne into the 21st Century*, Melbourne: Department of Planning and Development, Victorian Government.

DUAP (Department of Urban Affairs and Planning) (1998) *Shaping Our Cities: the Planning Strategy for the Greater Metropolitan Region of Sydney, Newcastle, Wollongong and the Central Coast*, Sydney: Department of Urban Affairs and Planning, NSW Government.

Forster, C. (2004) *Australian Cities: Continuity and Change*, 3rd edition, Melbourne: Oxford University Press.

Gleeson, B. and Low, N. (2000) *Australian Urban Planning: New Challenges, New Agenda*, Sydney: Allen & Unwin.

Greenwood, G. and Laverty, J. (1959) *Brisbane 1859–1959: A History of Local Government*, Brisbane: Brisbane City Council.

Hall, P. and Pain, K. (eds.) (2006) *The Polycentric Metropolis: Learning from Mega-City Regions in Europe*, London: Earthscan.

Hamnett, S. (2000) "The Late 1990s: Competitive Versus Sustainable Cities," in S. Hamnett and R. Freestone (eds.) *The Australian Metropolis: A Planning History*, St. Leonards NSW: Allen & Unwin, pp. 168–88.

Harris, C.P. (1978) *Local Government and Regionalism in Queensland, 1859 to 1977*, Canberra: The Australian National University.

Howe, R. (2000) "A New Paradigm: Planning and Reconstruction in the 1940s," in S. Hamnett and R. Freestone (eds.) *The Australian Metropolis: A Planning History*, St. Leonards NSW: Allen & Unwin, pp. 80–97.

Huxley, M. (2000) "Administrative Cordination, Urban Management and Strategic Planning in the 1970s," in S. Hamnett and R. Freestone (eds.) *The Australian Metropolis: A Planning History*, St. Leonards NSW: Allen & Unwin, pp. 131–48.

Lennon, M. (2000) "The Revival of Metropolitan Planning," in S. Hamnett and R. Freestone (eds.) *The Australian Metropolis: A Planning History*, St. Leonards NSW: Allen & Unwin, pp. 149–67.

McLoughlin, B. (1992) *Shaping Melbourne's Future? Town Planning, the State, and Civil Society*, Cambridge: Cambridge University Press.

McNabb, P. (2001) *Activity Centres Review: A Study of Policy and Centres of Activity in Metropolitan Melbourne and Geelong*, Melbourne: Peter McNabb & Associates.

MMBW (Melbourne Metropolitan Board of Works) (1954) *Melbourne Metropolitan Planning Scheme*, Melbourne: Melbourne Metropolitan Board of Works.

—— (1971) *Planning Policies for the Melbourne Metropolitan Region*, Melbourne: Melbourne Metropolitan Board of Works.

—— (1980) *Metropolitan Strategy*, Melbourne: Melbourne Metropolitan Board of Works.

MPE (Ministry for Planning and Environment) (1987) *Shaping Melbourne's Future: the Government's Metropolitan Policy*, Melbourne: Ministry for Planning and Environment, Victorian Government.

Morison, I. (2000) "The Corridor City: Planning for Growth in the 1960s," in S. Hamnett and R. Freestone (eds.) *The Australian Metropolis: A Planning History*, St. Leonards NSW: Allen & Unwin, pp. 113–30.

Newman, P. and Thornley, A. (2005) *Planning World Cities: Globalization and Urban Politics*, Basingstoke: Palgrave Macmillan.

O'Connor, K. (1993) *The Australian Capital City Report 1993*, Melbourne: Centre for Population and Urban Research, Monash University.

OUM (Office of Urban Management) (2005a) *South East Queensland Regional Plan 2005–2026*. Brisbane: Office of Urban Management, Queensland Government.

—— (2005b) *South East Queensland Regional Infrastructure Plan and Program 2005*, Brisbane: Office of Urban Management, Queensland Government.

Phares, D. (ed.) (2004) *Metropolitan Governance without Metropolitan Government*? Aldershot: Ashgate Publishing.

Proudfoot, H. (2000) "Founding Cities in Nineteenth-Century Australia," in S. Hamnett and R. Freestone (eds.) *The Australian Metropolis: A Planning History*, St. Leonards NSW: Allen & Unwin, pp. 11–26.

RCC (Regional Coordination Committee) (1995) *South East Queensland Regional Framework for Growth Management 1995*, Brisbane: Regional Coordination Committee, Queensland Government.

Ross, C. and Doyle, J. (2009) "The Megaregion and the Future of American Planning," in C. Ross (ed.) *Megaregions: Planning for Global Competitiveness*, Washington, DC: Island Press, pp. 280–87.

Salet, W. and Thornley, A. (2007) "Institutional Influences on the Integration of Multilevel Governance and Spatial Policy in European City-Regions," *Journal of Planning Education and Research*, 27(2): 188–98.

Searle, G. (1996) *Sydney as a Global City*, Sydney: Department of Urban Affairs and Planning.

—— (2004) "Planning Discourses and Sydney's Recent Metropolitan Strategies," *Urban Policy and Research*, 22(4): 367–91.

SPA (State Planning Authority) (1968) *Sydney Region Outline Plan 1970–2000 AD*, Sydney: State Planning Authority, NSW Government.

SSC (Smart State Council) (2007) *Smart Cities: Rethinking the City Centre*. Brisbane: Smart State Council, Queensland Government.

Spearritt, P. and DeMarco, C. (1988) *Planning Sydney's Future*, Sydney: Allen & Unwin.

UN Habitat (2008) *State of the World's Cities 2008/2009: Harmonious Cities*, London: Earthscan.

—— (2009) *Planning Sustainable Cities: Global Report on Human Settlements 2009*, London: Earthscan.

Yigitcanlar, T., O'Connor, K. and Westerman, C. (2008) "The Making of Knowledge Cities: Melbourne's Knowledge-based Urban Development Experience," *Cities*, 25(2): 63–72.

9 The Upper Spencer Gulf Common Purpose Group

A model of intra-regional cooperation for economic development

Jim Harvey and Brian Cheers

Introduction

In recent years, the global notion of regional development has changed, from an emphasis on targeted policies to redress the problems of uneven growth in lagging areas to a more positive understanding of the need for all regions to develop their competitiveness through value-adding in such a way that business can become more successful against both national and international competition.

In Australia, employment has become a key concern for both the government and regional communities. Unemployment rates have become a dominant political measure of the federal government's economic management. Therefore, regional development initiatives have been predominantly economic in nature and labor force in focus. Emphasis has been on short-term projects to increase the employment rate. This has created tension between the concerns of regions (particularly those in long-term decline and with high rates of unemployment) and federal governments.

Declining regions are characterized by a vicious cycle of welfare dependency, service withdrawal, depopulation, and deteriorating quality of life. Some policies are made to revitalize these regions. However, approaches that require all tiers of government to consult and lead at the local level appear to be out of step with the established neoliberal political structures and modes of operation. Consequently, the long-term mechanisms needed for breaking the cycles that entrap and isolate declining regions seldom exist.

This chapter reports on a regionally based initiative known as the "Upper Spencer Gulf Common Purpose Group" (CPG). The CPG was established in 1998 to ameliorate the rapidly deteriorating economy of the Upper Spencer Gulf Region in the state of South Australia by fostering cooperation and collaboration between the local governments, industries, and businesses of the three cities within the region. The region is one of contrasts: extensive rural and semi-desert areas in its interior, yet with three major urban areas hugging the coast of Spencer Gulf and around 100 kilometers apart. Approximately 80 percent of the population resides in the seaside centers that were developed to serve the traditional industries of the Upper Spencer Gulf: minerals extraction and processing; shipbuilding; steel and heavy manufacturing; and railways. The region is perhaps a "mega-region" more by area than by population size. From 1970 to 1998, these industries were seriously impacted by national and international changes to their markets and experienced considerable downsizing and restructuring. Previous attempts to strengthen the regional economy through government programs had failed to reverse the decline. This was due in part to the region's long history of intercity rivalry and insularity. The formation of the CPG represented an initially tentative move from separate and, in some ways, parochial and competitive attempts to tackle the economic, demographic, and social decline of the three cities to focus on collaboration as a vehicle for repositioning and reinvigorating its competitive advantages over a ten-year period.

In 2001, the Australian Department of Transport and Regional Services commissioned the Centre for Rural and Regional Development to review the CPG as an example of best practice in intra-regional cooperation involving two or more local governments. This chapter is based on the final report.

In this chapter, we describe the CPG, trace its development, and analyze the processes that underpin its focus and operation. From an analysis of similar structures in other parts of the world and regional development "best practice" principles, we identify a number of dimensions that can be used to describe organizations that are established to facilitate intra-regional cooperation. We present a set of key process performance indicators for such organizations that were developed during the study from which this chapter is derived, and use these to evaluate the CPG as a unique example of "best practice" in intra-regional cooperation.

The CPG model is unique. However, we were able to demonstrate that the long-term involvement of multiple stakeholders with all levels of government is clearly essential for overcoming previous failures, and those initiatives that are capable of doing this, and of remaining stable, deserve the label best practice. For these and other reasons the CPG, which is both stable and a bridge between local and central government, is a candidate for best practice.

Regional development in Australia

Australia is a federation of six states and two territories. While local government has no formal constitutional status in the Commonwealth of Australia, the effect of Australian federalism on regional development is that responsibility is divided across the three tiers of government: federal, state, and local. This has generated an approach to regional development that is fragmented, lacking in coordination, small-scale, and with a non-metropolitan focus, despite the fact that many metropolitan regions have become economically disadvantageous as a result of the current economic downturn. This has created tension between the center and the periphery, especially in regions experiencing decline.

In most states, the bulk of the funding and policy resources for intervening in regional economies rest centrally with the Federal Government, who has generally shied away from specific policies or projects. Regional development agencies have tended to respond to local regional issues with piecemeal and short-term initiatives using Commonwealth grants as their major source of funding. State governments in New South Wales, Western Australia, and South Australia have been substantial investors in regional development. Their principal input has been funding regional development boards across their non-metropolitan areas. Local government contribution varies greatly.

As Beer et al. (2005) succinctly state:

> Regional development agencies in Australia are in many ways a *product* of neo-liberalism, since they represent one way in which governments can be seen to be responding to regional pressure for assistance but they can do so without incurring significant costs.

This was the ideological and policy context in which the CPG was established.

The Common Purpose Group: bringing cooperation into regional development

In 1998, a comprehensive independent study by the South Australian Centre for Economic Studies concluded that the Upper Spencer Gulf cities were performing poorly compared with other provincial cities in South Australia. This study examined a number of social indicators such as population size, unemployment, and welfare dependency, and economic measures such as employment growth and household income. Along the lines of other studies, it reported downward trends for population size over the last two decades. At the time, there was no evidence to suggest that this generally downward trend would be reversed without significant government intervention. However, previous attempts to strengthen the regional economy through government programs had failed to reverse the decline.

The CPG was established in this context to face the economic and social realities confronting the region, and to attempt to reverse the decline and build a foundation for sustained long-term growth by fostering intra-regional cooperation. From its inception, the CPG has pursued this aim by: (1) acting as a peak economic development organization for the region; (2) engaging in strategic regional planning—positioning the region in state, national, and international markets; (3) facilitating economic development; (4) providing the region with access to information; and (5) building local business capacity. The focus has been for the CPG to become recognized by the major regionally based and external stakeholders as the legitimate cross-city economic development organization.

The CPG was established as an incorporated body, with a flat organizational structure that was to operate with a degree of formality at the committee level, but through extensive informal interaction outside of meetings. Core members were, and continue to be, the key regionally based stakeholders in regional development. Membership of the CPG comprises the mayors and chief executive officers (CEOs) of the three local governments, the CEOs of the three regional development boards, and a representative from each of the combined Chambers of Commerce and trade unions, the Center for Regional Engagement, University of South Australia, and the Spencer Institute of Technical and Further Education.

The CPG focus was to be clearly on economic (rather than social or community) development at the regional (rather than community) level, and was to be project (rather than politically or administrative) driven. Nevertheless, it had a dual focus on economic markets and the networks in which it was embedded. The CPG was expected to make good deliberate use of this embeddedness to maximize its network capability and its access to high levels of technological, intellectual, and social capital. Key elements of the CPG model are summarized in Table 9.1.

Principles of best practice in intra-regional cooperation

Nongovernment, nonprofit, regional development organizations such as the CPG exist to facilitate economic growth, population expansion, and increased export earnings and the like, within the region. However, such organizations are established not to achieve growth directly but rather to make certain processes happen, for example, to coordinate, facilitate, and foster cooperation among the key players so that the region can grow economically through increasing profits and export earnings.

Traditional evaluations of regional development initiatives tend to be driven by neoclassical economics. Consequently, they can be opaque to the qualitative subtleties of local politics, the interactions and relationships that make a regional economy tick, and the social, business, and government networks through which these happen.

From this more sociological perspective, the CPG is both constituted by and embedded in networks of social relations and interaction—both within the region and between the region and the outside world. Indeed, it owes its very existence to the fact that intra-regional cooperation happens through certain kinds of relations between individuals and organizations. In this sense, its ostensible purpose is to support and develop cooperative interaction for economic development. Social relations, interaction, and networks have a crucial role in the establishment, formation, and operation of the CPG.

There has been a change in the focus of regional development policies from targeted policies to redress the problems of uneven growth in areas that are lagging to a more positive understanding of the need for regions to develop their competitiveness through value-adding in such a way that business can become more successful against national and international competition. In non-metropolitan settings regional development programs are typically devised in an attempt to deal with problems of economic distress and social exclusion.

Best practice in regional development recognizes that these problems make up a complex of interlinked economic and social phenomena, and that finding solutions requires a multifaceted approach focusing not only on issues of investment and production but also on human attitudes, relations, and networks. Effective regional development, while encompassing either the immediate locality (or community) and the wider region, distinguishes between these levels. The CPG took the view that analysis of regional development should focus on the regional, or multi-community level, while also recognizing that certain tasks that are integral to the development of a region are best planned and executed at the local level.

A central generating principle of effective regional development is communication between diverse actors, both within the region and beyond it. Best practice in regional development provides a field of dialogue between development ideas and initiatives originating at the state, regional, and community levels. By breaking down the compartmentalization that isolates policy issues within separate agencies, programs, and services, best practice in regional development fosters new strategies and policy innovations and encourages new alliances. Familiarizing participants with the organizational capacities of other partners, it multiplies the potential for integrated, or at least coordinated, program development and implementation.

There are good reasons for taking a social angle on regional development. Community development research has consistently found a positive relationship between community activeness (in the sense of citizen involvement and the commitment of time and resources to activities for the good of the entire community) and a range of desirable development outcomes, such as economic growth, higher employment and income levels, and successful resolution of social problems (Wilkinson 1991; Bourke et al. 1996; Luloff 1996, 1998; Claude et al. 1999; Cheers and Luloff 2001). Indeed, studies indicate that economic growth is more likely to be the product of community development than its cause (e.g. Humphrey and Krannich 1980).

Clearly, if regional development efforts are to maximize their impact, they need to rest on a sound base of localized community development. It follows, then, that best practice in regional development includes helping local communities to identify promising projects and secure state support. The clear message from the literature is that regional development efforts that fail to link up with the conditions, aspirations, needs, leaders, and structures at the community level are less effective than those that do.

This angle on regions and regional development emphasizes partnerships, networks, and systems within and outside the region. Common threads in this literature include the need for the participation of a broad base of stakeholders in the development process, and the importance of balancing power relationships both within the region and between the region and external stakeholders.

Table 9.1 Key elements of the CPG model

Descriptive dimension	*Element*
Structural formality	Incorporation Structural independence A flat organizational structure Relative overall informality Varying formality at different levels: • Management committee—relatively formal; • Implementation team—relatively informal; • Member interaction outside meetings—informal
Funding (financial and in-kind support)	Member commitment Other intra-regional commitment Central government commitment, focused on particular areas of activity and providing some temporal security
Focus	Region
Market/network mindset	Market transactions Networks within and outside the region
Sector focus	Economic development, including tertiary education
Roles	Peak regional economic development organization Facilitating cooperation and partnership development Engaging in and facilitating strategic, region-wide economic development Market-positioning Facilitating economic development Providing regional access to information and facilitating information exchange Building local business capacity
Intra-regional partnership development	Providing a structural location for cooperation Facilitating the development of shared agendas Facilitating communication and information flows Mediating intra-regional relationships Creating new partnerships Mobilizing existing partnerships Managing competing local interests and tensions

Table 9.1 Continued

Descriptive dimension	*Element*
Stakeholder representation in management structures	Core members include: • political and administrative arms of local government; • key economic and/or regional development organizations; • the business sector; • trade unions; • tertiary education institutions; • directly related central government departments (observer status)
Locus of initiative	Local
Operational autonomy	Structural and functional autonomy
Locus of control	Local Subsidiary, supportive role for central government
Locus of intra-regional power	Mostly fairly egalitarian Local government and EDB representatives most central Floating power distribution, depending on the issue and project
Network capability	Horizontal embeddedness in local governance, professional and business networks Vertical embeddedness in central government networks
Legitimacy	Within the region Outside the region
Access to social, technological, and intellectual capital	Trust, reciprocity, and cooperativeness in relationships and interaction: • within the common purpose group; • between members and external people and organizations; • in the group's environment Technological capital—infrastructure, including databases Intellectual capital—professional expertise relating to regional and economic development, and particular projects

Source: Cheers et al. 2002.

Best practice in regional development coordinates the energies of a wide range of actors in the economic and social sectors, rather than simply enlisting the support of particular regional elites.

Community development, partnership, network, and systems approaches all emphasize the level of the community or network. However, in Australia, the bulk of funding and policy resources for intervening in regional economies rests centrally with the Commonwealth. Hence, there frequently exists tension between center and periphery regions in achieving a "top-down-plus-bottom-up" consensus for development.

Almost by definition, regions in decline are beset by market failures and breakdowns in institutional and community cooperation. Furthermore, a substantial component of regional structural adjustment policy has been aimed at correcting the results of previous policy errors (Mortimer Report 1997). A vicious cycle of welfare dependency, service withdrawal, depopulation, and deteriorating quality of life sets in. This cycle hampers formation of the dynamic networks and the broad participation that characterizes effective regional development, and this, in turn, exacerbates the decline. Policies based on the inherent rightness of market forces are not always sympathetic to programs that can be justified only in terms of the long-term social value of a community or on the idea of market failure, while the short-termism often pervades development policy. In situations of long-term deterioration, therefore, policies limited entirely to influencing market forces in the short term—economic "quick fixes," or band aids—tend to be ineffective. In contrast, policies aiming at more fundamental economic and social reconstruction require long-term political will, along with a commitment to multifaceted intervention both in the regional economy and in the relations between industries, businesses, organizations, government (at all levels), and people through which economic transactions are made.

The long-term involvement of multiple stakeholders with all levels of government is clearly essential for overcoming these failures. Initiatives that are capable of this, and of remaining stable, deserve the label "best practice."

The role of cooperation between local authorities in best-practice regional development

Local governments are the cornerstone of regional development in Australia and have generally been the level of government most successful at creating stable and durable regional development organizations (Beer and Maude 1997). Local governments have important strengths as development agents. They are keenly aware of local conditions and dilemmas, and they are embedded in local information flows and networks of influence. Above all else, their status as elected bodies with statutory responsibilities lends legitimacy to development coalitions that would otherwise be perceived as answering only to local or external elites.

Nevertheless, as exclusive managers of development initiatives, local governments are less than ideal. For instance, Beer and Maude (1997) found that many local governments are not well placed to be lead agencies in regional development because they are insufficiently representative, lack the required professional expertise, lack resources, and are too small and parochial. Major decisions by local councils are often the outcome of a complex system of political ploy and counter-ploy through which local interests weigh up and trade off their likely gains and losses. As a consequence, decisions tend not to be taken quickly, and bold strategic thinking is the exception rather than the rule. At worst, local government processes can allow sectional interests to block, quite undemocratically, moves that would win wide acceptance if they were submitted to broader local forums.

Additionally, local governments frequently lack the professional expertise required to be successful development agents. Elected representatives and employed staff rarely have the required training in the development field, and modest funding bases mean that small councils, in particular, find it financially burdensome to hire consultants.

On the basis of such shortcomings, Curran and Blackburn (1994) concluded that locally based organizations, such as local governments and chambers of commerce, are steadily becoming less capable of affecting industrial and economic change within declining and delocalizing industrial economies.

The most useful role of local authorities in development initiatives appears to be as partners in dedicated coalitions rather than as the prime movers. In fact, the role of coalition partners is likely to suit local governments, because this: enables them to keep development and local maintenance functions separate; reduces their responsibility for development failures while allowing them to take some credit for successes; protects their image as sound economic managers from the intrusions of more messy social development; and isolates them from legal and financial risk.

There are many development functions, especially economic ones, that tend to fit poorly within the boundaries of a single local authority. First, economic strategies are often better devised and implemented on a regional, rather than local, scale because the pooling of resources can allow local authorities to afford the expertise, technology, and access to data that development requires. Second, encouraging businesses and industries to cluster so as to share technological skills and achieve better access to markets and relevant information will often be more practical on a regional scale than within a specific local authority area. Third, regional partnerships allow the marshalling of resources needed to meet large-scale infrastructure needs. Fourth, regions are frequently better able than individual local authorities to attract major investment from external sources, provided that structures exist for managing potential intra-regional competition. Fifth, advocacy on a regional basis provides more political pull with Commonwealth and state governments. Finally, as observed by Cashin (1999) in a widely quoted American study, regional development organizations can bring together a large pool of decision-makers representing a wide cross-section of interests. This can offset the danger of sectional interests and pressure groups seizing power at the local level, resulting in the pursuit of minority interests to the detriment of the economic and social development of the region as a whole.

Purpose-built coalitions are not the only instruments that development-minded interests may set in place at the regional level. An alternative form is the "super council," involving the ceding by local authorities of various regulatory and development functions to a higher, regional tier of government as in Minneapolis-St. Paul and Portland, Oregon (Cashin 1999). Although "super councils" have been shown to have some strength, they suffer from many of the same shortcomings as local government bodies. For example, cooptation by special interests may still occur. "Super councils" also add yet another layer of statutory bureaucracy for development practitioners, investors, and markets to deal with, and can be unresponsive to changing market conditions and opportunities.

The need for regional development coalitions to relate actively to central government as an important stakeholder can be accepted as a further principle of best practice. Nevertheless, central government involvement in regional development organizations inevitably gives rise to tensions that must be handled with tact and diplomacy. The extensive research on social capital has shown that effective economic and social development is closely associated with relations characterized by familiarity, trust, mutuality, and reciprocity. Government representatives on development coalitions, however, speak for government departments that have the power to approve or deny much-needed funding. The possibility follows that the relationship between government representatives and coalitions will not be one of reciprocity, but of donor and supplicant. Governments also have particular agendas that their representatives, especially in pre-election times, may be tempted to try to impose. The point is that in the best practice of regional development, central government is a partner, not a dominating presence. Regional development coalitions need to have an independent existence, answering to regional aspirations rather than simply carrying out government policy. The eighteen principles for effective intra-regional cooperation we have identified are summarized as below:

1. recognize the complexity and interrelatedness of regional economic and social development issues;
2. focus on issues of investment and production, as well as the social networks and relations in which these are embedded;
3. involve stakeholders across sectors within the region;
4. involve all tiers of government;
5. engage intra-regional, extra-regional, and government stakeholders with each other;
6. promote communication and interchange between diverse sectors, to create links between the development of ideas and initiatives originating from stakeholders;

7. relate top-down leadership to bottom-up participation;
8. develop a broad and stable political base to offset domination by particular interest groups;
9. develop cooperation between local authorities as members of dedicated coalitions, rather than as the prime movers of regional development.
10. emphasize regionally based development;
11. differentiate and rationalize interaction between regional and community layers in development;
12. ensure that central governments act as partners, not as a dominating presence;
13. ensure that adequate and predictable funding is provided, independent of electoral cycles, which provides for stability and effective planning;
14. provide multi-track dialogs and feedback between the cooperative regional development organization and industry, community partners, and government;
15. insulate cooperative regional development organizations from excessive bureaucracy;
16. employ realistic appraisals of regional capabilities, technology cycles, and competition;
17. provide access to expert advisors and best practice knowledge; and
18. help local communities to identify and secure investment and funding for promising projects.

These, then, are some key best-practice principles derived from regional development literature for intra-regional cooperation in situations where local government is an important player. What can we learn from reviewing actual attempts at establishing such structures?

Models of cooperation

Collaboration between local authorities is not unusual in regional development. However, this has mostly been limited to improving efficiency and cutting costs in traditional local government functions, rather than as a genuine regional development tool.

Regional economic development started gaining legitimacy as an accepted pursuit for local authorities in various parts of the world from the 1970s, as concepts of the "entrepreneurial city" took over from the earlier "managerial" paradigm (Harvey 1989). For local governments to grasp the desirability of their collaborating on development tasks, then, a further round of experiment and reflection was required. The result has been that a significant range of models of such collaboration has begun to appear only in the last few years.

In some countries, national circumstances and traditions have militated against local governments setting out to collaborate on regional development tasks. In the US, for example, the tendency has been for the private sector, rather than the public sector at any level, to provide the initiative, leadership, and a good deal of the funding for development efforts.

In Europe, regional development has been assigned high priority by the European Commission and, in most cases, by national governments. Programs of considerable sophistication have been devised and, for the most part, have been overseen by well-resourced government apparatuses. However, the main nexus of ideas, power, and funding has tended to be state–local, with regional elements less prominent. So, to the extent that local governments in Europe collaborate on regional development tasks, this is generally mediated by state development agencies, rather than through regional stakeholder organizations created for this purpose. Furthermore, some European countries have intermediate tiers of government that assume particular regional development functions. In sum, it is quite rare for local authorities in Europe to engage in direct, structured, and extensive collaboration in regional development.

As indicated earlier, local governments are not the ideal bodies to devise and implement regional development strategies. Where the need for regional development is acute, however, local authorities are among the first government bodies to feel the lack of effective programs. The ebbing of local tax revenues as populations and property values decline places a range of local authority operations in peril. With their ties to local small businesses, local governments come under strong political pressure to do something to raise stagnant or declining profits. The result has often been to make local authorities especially prominent among the bodies calling for structured development programs.

Among local governments in Australia, a sense of neglect by higher levels of administration is particularly widespread. At the federal level, the interest shown in regional development has been inconsistent. From 1994, the Keating Labor government made a significant rhetorical commitment to support the regions. In 1996, this position was reversed by the new coalition government, which argued that regional development was properly the concern of the states. Never has federal funding available for regional development exceeded what has been provided by state and local governments (Beer and Maude 1997).

It is salutary to note that there has been a significant increase in funding for regional infrastructure development as a result of the federal government's responses to the impact of the current global financial crisis. Targeted funding for schools, hospitals, roads, railways, water supply, and IT infrastructure has increased significantly. For example, the Building the Education Revolution (BER) Fund is a $16.2 billion investment over three financial years (2008–11), and is a key element in the Rudd government's $42 billion Nation Building and Jobs Plan. However, the allocation of funds has tended to be driven by national, rather than regional priorities and *not* by regional development agencies. Priorities and specific allocations have been determined at a national level, and the funds have been dispersed from central agencies and sometimes through states, but rarely through local government agencies.

Australian state governments, meanwhile, have only modest funding resources, and in two of the states where the need for regional development is most acute—Tasmania and South Australia—chronic near-recession has meant that the possibility of using state-derived revenue sources to fund development is especially slim.

Therefore local governments in depressed areas of Australia have a strong incentive to take up the tasks of economic and social development themselves. With neighboring local authorities facing many of the same dilemmas, joint action is an idea that, at a certain point, must force itself on even the most parochial of local administrations. But for local councils to carry on intensive, structured collaboration is not easy. As well as having to cope with unfamiliar tasks, local governments that embark on this course may have to overcome traditional rivalries with neighboring bodies. Launching regional development efforts comes at a definite cost, and, at least in the short term, a good deal of what is spent must be paid for from elsewhere in the regional economy. Councils must also surmount the fear (and sometimes the definite prospect) that even if costs are equally shared, the benefits of development will be concentrated in particular geographic centers within the region. Finally, local governments that set out to collaborate on regional development face the problem that, while there exists a need to harness local energies to build partnerships with a pronounced bottom-up element, they are generally doing so in regions without a strong culture of community development.

Towards a performance evaluation model of best practice in regional development

Assessment of principles and key performance indicators (KPIs) of best practice in regional development is both a complex and an underdeveloped area. There are typically two main approaches: (1) assessment against "hard" outcome indicators (e.g. how many jobs have been created); and (2) various forms of formative assessments of practice (e.g. how well the regional development organization undertakes certain processes). Both forms of assessment are limited because many of the social, economic, and political phenomena associated with regional development have both their origins and their solutions in arenas that are beyond the control of those involved in developing a particular region.

Our CPG report focused more on principles and indicators relating to processes rather than outcomes, and it was considered inappropriate to evaluate CPG outcomes at that time (2002). It was too soon to assess any impact on the vision of turning around the economic and population decline by 2010. Also, the usual indicators, such as export earnings, unemployment levels, and population growth, could not be directly attributed to the CPG.

Our initial intention was to identify a way of describing the CPG to enable us to compare it with other similar organizations. We had initially hoped to develop a comprehensive, process-focused performance evaluation model

(PEM) from a case study of one organization. We discovered that this was a much larger task than we had anticipated. (For a detailed discussion of the development of three sets of process KPIs, see Cheers et al. 2002: 21–23.)

Our starting point was the development of the eighteen principles of best practice in intra-regional cooperation. These were generated by analyzing Australian and international case studies. In theory, we should be able to consolidate the data into a single framework. However, this would be premature because the work of developing a coherent theoretical base has not been done, different conceptual levels are evident, indicators are far from exhaustive, our data are predominantly qualitative, and indicators have not yet been validated.

Review of the CPG against indicators of best practice

We now turn to an evaluation of the performance of the CPG against the principles of best practice generated above and specified in the previous section. For the sake of brevity, we have collapsed our review to an integrated list of indicators:

Indicator 1: Recognize the complexity and interrelatedness of regional economic and social development issues. Documentation and interview narratives concerning the formation, establishment, and development of the CPG reveal clear recognition among the organization's founders and other members of the challenges and complexities involved in reversing the long-term serious downward spiral of the regional economy. This recognition was born out of the many failed previous attempts at government intervention, the analysis contained in the report of the South Australian Centre for Economic Studies (1998) that was the initial trigger for the CPG's formation, and lessons learned from other similar situations. Recognition of the complexity of regional development issues is demonstrated in the organization's strategic plan, its multi-pronged attack on the problem, and its willingness to seek the expertise of a wide range of stakeholders, including all levels of government, a range of government departments, a variety of professional regional developers, and various databases.

Members are well aware of the interrelatedness of economic and social development. Nevertheless, because the CPG has not focused on social and community development, it has little access to information and expertise concerning these. Should the organization extend its focus to include, for example, health and human services, it would need to expand its membership base, networks, and access to expertise.

Indicator 2: Focus on issues of investment and production as well as the social networks and relations in which these are embedded. The CPG is focused squarely on the economic development of, and enterprise development within, the region. Nevertheless, it is also well embedded in local business and political networks and relevant extra-regional networks. The CPG gives high priority to engagement with stakeholders within the region and with relevant government departments, partners, and markets outside it. The organization originated, and remains firmly embedded, in sound social networks and relations among members and other key stakeholders. The importance given to these by participants in the study was somewhat surprising given their lack of formal community development experience and expertise, or sociological training.

Indicator 3: Involve stakeholders across sectors within the region. All key stakeholders and sectors involved in regional economic development are represented within the CPG. A variety of activities (e.g. industry clusters, customer-supplier forums, delegations, workshops, seminars, and lectures) facilitate the interchange of ideas and initiatives.

Indicator 4: Involve all tiers of government. All tiers of government are involved in the CPG structure and activities.

Indicator 5: Engage intra-regional, extra-regional and government stakeholders with each other. The CPG actively engages in all the roles identified from regional development literature and case studies (see Cheers et al. 2002: Chapter 4).

Indicator 6: Partnership development. The CPG prioritizes all the roles identified in relation to partnership development. Within the region, the focus is on partnership development among: (1) enterprises, (2) economic development, political, education, business, and workforce sectors; and (3) regional development professionals and politicians.

Indicator 7: Relate top-down leadership to bottom-up participation. An effective balance between leadership and participation is obtained through the membership structure and processes of the CPG. Local authorities provide a bottom-up connection with local communities, along with other local public-sector partners, including universities and training organizations. Employers, trade unions, government departments, and agencies are represented. This structure and associated processes facilitates positive outcomes for the CPG.

Indicator 8: Develop a broad and stable political base. The CPG clearly has a broad and stable political base, which has successfully managed any potential for domination by particular power blocs. This is built on strong commitment to cooperation and reasonably equal representation by key relevant stakeholders.

Indicator 9: Develop cooperation between local authorities as members of dedicated coalitions. The Port Augusta, Port Pirie, and Whyalla Councils certainly cooperate as members of a dedicated coalition that is, collectively, the prime mover of regional development. Together with their associated Regional Development Boards, they are the most influential members of the CPG, which is appropriate given their leadership roles in their respective cities, and the combination of political and professional legitimacy and expertise they bring to the organization.

Indicator 10: Emphasize regionally based development. The CPG focuses on the development of the region as a whole. Its links with local communities are through organizations (Regional Development Boards (RDBs) and local councils), rather than with more "grassroots" community development organizations. Nevertheless, within the business sector, the CPG is well connected with this level through links with individual enterprises, which are grouped by industry rather than cities. This is appropriate and effective, given the organization's aims.

Indicator 11: Differentiate and rationalize interaction between the regional and community layers in development. The CPG has differentiation as a primary objective of the organization. This is driven by the realization that in order for the region to survive, competitive advantages have to be created in order to differentiate the region from others. A variety of strategies have been employed to provide a better investment environment to ensure the region is at maximum attractiveness to potential investors.

Indicator 12: Central governments act as partners, not as a dominating presence. The CPG is locally driven, controlled, owned, and self-determining. Involvement of central government is very clearly targeted and appropriate to the organization's strategic plan. State and Commonwealth governments support, rather than dominate and direct, the organization.

Indicator 13: Provide adequate and predictable funding. CPG funding is adequate, although not so generous as to encourage complacent, unfocused, or peripheral activity. It is carefully targeted at the priorities identified in the strategic plan. While funding is secure enough to allow planning and reasonable stability, it is not entirely independent of electoral cycles. This is probably inevitable.

Indicator 14: Provide multi-track dialogues and feedback between the cooperative regional development organizations and industry, community partners and government. Overall, the CPG is highly responsive to its environment. Dialogue with, and feedback from the RBDs, the three local governments, and the business, economic and union sectors is ongoing. Links are also strong with the relevant state and Commonwealth Government departments, because these departments are directly involved in the CPG through their observer status and they fund three key positions. Dialogue is also strong with local industry through the CPG's relationships with individual enterprises, the Resource Industry Cluster, and direct involvement of the combined Chambers of Commerce and trade unions. As suggested earlier, links with other community organizations are not evident, although this is not important as long as the CPG remains focused on economic and enterprise development.

Indicator 15: Insulation from excessive bureaucracy. The CPG is well insulated from excessive bureaucracy because its resources are too limited to support a bureaucracy, member organizations maintain their spheres of operation independently of the CPG, the organizational structure is flat, and the CPG is separated from its working groups.

Indicator 16: Realistic appraisal of regional capabilities. The CPG was formed on the basis of a starkly realistic appraisal of the economic and industry environment within the region. This realism has been retained throughout its development. Its assessment of regional capabilities is based on a number of sources, including: ongoing interaction with regionally based enterprises; strong links with all sectors and stakeholders; access to a range of databases through the RBDs; the local, state, and Commonwealth governments; the University of South Australia and the Spencer Institute of Technical and Further Education; and various consultancies. The growth in the number of export-ready and export-effective enterprises within the region and the number of contracts won by local enterprises during the construction of the Adelaide to Darwin railway indicates that the CPG's assessment of capability has been accurate.

Indicator 17: Provide access to expert advisors and best practice knowledge. The CPG has employed private consultants and utilized the expertise and resources resident in the Center for Regional Engagement, University of South Australia, Whyalla. The consultants have provided position papers on topics such as industry clusters, together with workshops and one-on-one assistance to regional industries to enable them to be both quality assured and export-ready companies. The latter have been pre-requisites for pre-registration for major regional infrastructure project tenders. The university has provided seminars, workshops, research design, and statistical analysis and collaborative partnerships for industry-based research and development grants.

Indicator 18: Helping local communities to identify promising projects and to secure investment funding for them. This has been an overriding focus of the CPG. If we regard the three cities as the "communities" in the region, the CPG has certainly supported these in the ways indicated by this principle. A number of projects have been won because all CPG members threw their collective weight behind the particular city's bid. Furthermore, in keeping with the CPG's strategic plan, "community" can also be defined here in terms of "communities of businesses," such as partnerships and industry clusters. The CPG has been highly supportive of these. The strategy of facilitating partnerships among local enterprises and helping them to assess their own capability and upgrade it has resulted in growth in exports from, and contracts flowing into, the region.

Conclusion and implications for other regions

The mission of the CPG is to turn around the economy and population decline by 2010. The Upper Spencer Gulf Common Purpose Group seems to be achieving its aim of fostering intra-regional cooperation for economic development extremely well. It has, in effect, turned around the relationships between the three cities with respect to economic development from insularity and rivalry to cooperation balanced by the principle of reasonable competitiveness. According to these process indicators, which arguably represent the true measures of success of the CPG, at least at this stage, it appears to have done well.

There are other indicators of substantial progress as the CPG develops initiatives aimed at turning around the economy and making local enterprises more competitive. Gaps in business capability and capacity have been identified and programs developed to fill these gaps. The appointments of an enterprise development officer and an export development officer has enabled some local companies to win contracts for work on the Adelaide to Darwin railway, either by themselves or as joint partners. Another outcome has been the broader vision that is now being demonstrated by some of the local enterprises with respect to their engagement in projects external to the region.

Invest Australia, the country's national investment agency, was commissioned to provide advice on investment opportunities in the Upper Spencer Gulf. The resulting report, *Making a Case: Investment in the Upper Spencer Gulf* (Invest Australia 2000), identified some specific investment opportunities in the region and provided a basis for community input to the development of a comprehensive regional strategy. An industry cluster based on the main industry in the region-resource processing was established in 2001. This attempts to capitalize on accumulated knowledge and networks residing among small and medium-sized suppliers to the existing regional resource processors to form a solid base to facilitate entry into global markets. This initiative has become more significant as massive new resource processing projects, such as the extension of the Roxby Downs mine to become one of the largest mines in the world, have been announced and other large-scale mines are being developed in the wider region. These and other initiatives indicate that the CPG has made significant ground in several areas, including: regional research, cluster formation, export culture, and regional infrastructure prioritization.

In conclusion, we suggest that the Upper Spencer Gulf Common Purpose Group presents a model of best practice in intra-regional cooperation in situations where two or more local governments are key players.

Transferability of the CPG model to other regions

Similar regions considering developing an organization with related goals would do well to study the CPG model in detail, with a view to adapting it to local conditions. The key elements of the CPG model are summarized in Table 9.1. Our analysis of a number of similar organizations indicates that these elements appear to be transferable to other contexts.

We suggest that elements of the CPG model are readily transferable to regions with the following characteristics:

- significant economic threat;
- a manageable geographic, demographic, and industrial/economic context;
- two or more local government areas that are either adjacent and/or readily geographically accessible to each other;
- sufficient determination to cooperate;
- a reasonable level of goodwill among most of the major stakeholders;
- current, or at the very least, emerging economic opportunities;
- a reasonable level of local business capability and infrastructure;
- central government support for the development of the region through locally driven intra-regional cooperation;
- access to moderate funds.

On the other hand, some elements, or the entire model, may not be transferable to, or may need significant modification in, other kinds of regions.

A number of elements of the CPG model have been crucial to its success at fostering economic development through intra-regional cooperation. Therefore, with the foregoing proviso, we recommend that they be seriously considered for adoption, with appropriate adjustments, by other regions developing similar organizations.

Finally, the process indicators and KPIs generated in this study and the framework for analyzing such organizations should be further developed for research and evaluation applications and for monitoring their progress by funding bodies, regional development researchers, and the organizations themselves.

References

Beer, A. and Maude, A. (1997) *Effectiveness of State Frameworks for Local Economic Development*, Adelaide: Southern Development Board Adelaide, the Economic Development Authority, and the Local Government Association of South Australia.

Beer, A., Clower, T., Haughtow, G., and Maude, A. (2005) "Neoliberalism and the Institutions for Regional Development in Australia," *Geographical Research* 43(1): 49–58.

Bourke, L., Jacob, S., and Pitman, S. (1996) "Response to Pennsylvania's Agricultural Preservation Program," *Rural Sociology*, 61: 206–29.

Cashin, S.D. (1999) *Localism, Self-Interest and the Tyranny of the Favored Quarter: Addressing the Barriers to the New Regionalism*, Georgetown University Law Center, Working Paper Series in Law in Business and Economics, Working Paper no. 194751. Online at http://papers.ssrn.com/paper.taf?abstract_id=194751 (accessed June 1, 2002).

Cheers, B. and Luloff, A.E. (2001) "Rural Community Development," in S. Lockie and L. Bourke (eds.) *Rurality Bites: The Social and Environmental Transformation of Rural Australia*, Annandale: Pluto Press, pp. 129–42.

Cheers, B., Harvey, J., Clarke, R., and Heffernan, P. (2002) *The Upper Spencer Gulf Common Purpose Group: A Model of Intra-regional Cooperation for Economic Development*, Whyalla: Centre for Rural and Regional Development, University of South Australia.

Claude, L., Bridger, J.C., and Luloff, A.E. (1999) "Community Well-being and Local Activeness", in P. Schaeffer and S. Loveridge (eds.) *Small Town and Rural Economic Development: Case Studies*, Westport, CT: Praeger, pp. 39–46.

Curran, J. and Blackburn, R.A. (1994) *Small Firms and Local Economic Networks: The Death of Local Economy?*, Chapman, London.

Harvey, D. (1989) "From Managerialism to Entrepreneurialism: The Transformation in Urban Governance in Late Capitalism," *Georgrafiska Annaler*, 71B(1): 3–17.

Humphrey, C.R. and Krannich, R.S. (1980) "The Promotion of Growth in Small Urban Places and its Impact on Population Change, 1975–78," *Social Science Quarterly*, 61(3/4): 581–94.

Invest Australia (2000) *Making a Case: Investment Potential in the Upper Spencer Gulf*, Canberra: Invest Australia.

Luloff, A.E. (1996) "The Doing of Rural Community Development Research," in H. Echelberger (ed.) *Rural America: A Living Tapestry*, Proceedings of the Research Fit Module, Third Annual US Forest Service, Radnor, PA: United States Department of Agriculture Forest Service, pp. 25–30.

—— (1998) *What Makes a Place a Community?* The Fifth Sir John Quick Bendigo Lecture, Bendigo: La Trobe University.

Mortimer Report (1997) *Going for Growth: Business Programmes for Investment, Innovation and Export*, Mortimer Report to the Minister for Industry, Science and Tourism, Canberra: Australian Government Publishing Service. Online at http://catalogue.nla.gov.au/Record/56995 (accessed December 15, 2009).

South Australian Centre for Economic Studies (1998) *Building an Economic Community: A Turnaround Strategy for the Upper Spencer Gulf Region to 2010*, Adelaide: SA Centre for Economic Studies.

Wilkinson, K.P. (1991) *The Community in Rural America*, New York: Greenwood Press.

Part III

State-led governance and planning under transition

10 Coordinating the fragmented mega-city regions in China

State reconstruction and regional strategic planning

Jiang Xu and Anthony G.O. Yeh

Introduction

Since the adoption of economic reform policy in 1978, several trends have been changing Chinese society, most significantly the rapid transition to an urbanized country. Not only is more than one-third of the country's population now living in cities but the remaining population is also becoming increasingly dependent on cities and towns for its economic survival and livelihood. At the National People's Congress in March 2001, the central government clearly affirmed the coming reality of China as an urban nation. This is further taken on by making urbanization an important national strategy in the Tenth Five-Year-Plan (2001–05). At the Sixteenth National Congress of the Communist Party of China held in November 2002, the Party reiterated its resolution to build an all-around *Xiaokang*—literally a moderately well-off society—of which a higher level of urbanization is a major component.

Along with this central policy, China has entered a period of fast growth in urbanization. Chinese experts predict that by 2050, the urban population is likely to reach around 1 billion, with the urbanization level soaring up to over 75 percent and the urban sector contributing over 95 percent of the national economy (Li 2003: 11). Accordingly, more than 600 million Chinese people will shift from rural areas to urban districts by 2050. Furthermore, projections show that by 2050, there are likely to be fifty ultra-large cities with populations of more than 2 million, over 150 big cities with populations of 1 to 2 million, 500 medium-sized cities, and 1,500 small cities (*People's Daily* 2002). While these figures are predictable, other forces such as globalization and mega-city region development will reinforce the role of cities and their regions as centers of production, consumption, and social and political transformation.

Urbanization at such a scale and speed has overwhelmed Chinese governments at various levels. The flood of rural–urban migration has exacerbated the infrastructure burden of cities, and led to the tremendous growth of un-serviced urban areas where millions of migrant workers lack adequate access to basic urban services. There is also a pressing need to address the problems of widespread misuse of land, urban sprawl, traffic congestion, and poor sanitation and living environments in all cities, especially those that are threatened by rapid and often uncontrolled growth, inadequate and poorly maintained infrastructure, industrialization, and increasing vehicle ownership. To effectively address these problems, China is redefining its urbanization strategy. Previously, small towns were emphasized as the main focal points for absorbing an excess of rural laborers, while large cities were under controlled expansion. The new urbanization strategy stresses the coordinated development of metropolises, medium-sized and small cities, and towns. Small urban settlements are allowed to grow as long as they can meet the required conditions. At the same time, there is a focus on developing urban agglomeration to replace the rampant urban sprawl (China City Mayor Association 2004). It is argued that urbanization should be primarily facilitated through the development of urban agglomeration. This brings the question of development and governance of mega-city regions to the fore.

Indeed, regions in China have undergone a significant transformation because of market reform, globalization, and rapid urbanization in the past two decades. Many cities and towns that were formerly peripheral or rural areas have developed into active economic centers linked to world capital investors and consumer markets. The resultant polycentric spatial form has combined with the rise of urban entrepreneurialism that has increasingly become a key municipal strategy to enhance place-specific socio-economic assets. This reform-imposed transition leads to intensified intercity competition for mobile capital.

There are a number of negative externalities that arise from this type of regional formation, such as widespread imitation, redundant construction, and environmental degradation. These externalities are profound in their impact, and different solutions are proposed in response. Mainstream literature on urban economics supports the idea of using competition as the way to sustain growth, because hyper-competition leads to economic efficiency, which may automatically solve the problems of collective action (Zhao 2002), and, thereby, state strategic concerns should be minimized. Whatever the intellectual shortcomings and political dangers of this mindset, it appears to be exercising a considerable influence upon the ideology, practice, and institutional infrastructure of urban and regional governance in contemporary China (Xu 2008). In contrast to these views, there is wide concern about the negative impact of political fragmentation. Commentators argue that strategic plans should be made to reduce this impact (Liu 2001). The motivation for this concern is varied, but objectives have typically been to reposition regions in the national and global economic landscapes as well as to articulate a coordinative spatial mosaic for key development issues such as land use regulation, environmental protection, and infrastructural provision. However, given the prevailing views that exhort cities to compete against one another, planning coordination for mega-city regions remains one of the most neglected policy areas in China.

Against this backdrop, this chapter focuses on the development and governance in the Pearl River Delta (PRD or delta thereafter). The PRD is a mega-city region in Guangdong, a coastal province in southern China. It covers nine municipalities with a resident population of more than 50 million. A former Maoist backwater, since economic reform the PRD has been transformed into a renowned world-manufacturing workshop and a prodigious exporter. A polycentric region, it pioneers the search for new territorial visions to coordinate regional growth. However, the difficulties associated with regional coordination are substantial. They arise from the very distinctive social and political context in a transitional economy. The case can therefore offer a good example for understanding the growth and management of mega-city regions when China is transformed from a centrally planned economy to a market-oriented economy.

This chapter is divided into six sections. The second section—which follows this introduction—reviews the development of mega-city regions in China and interprets the reasons why interest is surfacing in relation to this level/scale of development and planning. The third section briefly examines the development history of the PRD and how the delta has evolved into a fragmented region. The existing economic structure of nine delta cities are discussed to identify whether or not there is any kind of functional distribution among them. The fourth section investigates interactions that are taking place between the component cities of the PRD using the outcomes of a mapping analysis on the basis of available commuter data. We intend to answer how and to what extent delta cities are connected with one another in addition to their competitive relationship. In the fifth section we explore the implications for governance and planning. Existing institution and spatial policies are reviewed. The chapter ends with observations for theoretical discussion and policy implications.

Why plan for mega-city regions in China?

Regional development in China carries strong legacies of its socialist history. Under state socialism, horizontal linkages among jurisdictions were not considered important, with hierarchical linkage tending to predominate (see Chapter 1 of this volume). This resulted in regions being dependent on the center. Networked cities hardly existed. Regions were created by the central state to serve two primary purposes. One was to clear up the remnants of old regimes, while the other was to enforce central control (Solinger 1977). Urban and regional services were sectorally organized (Wu 2006).

After 1978, the transition towards a more market-oriented economy created new conditions for city networking. In the 1980s, there was a top-down interest in establishing economic coordination regions in order to explore horizontal linkages of central cities with their surrounding jurisdictions. More than 100 economic regions were then established. Examples include the Shanghai Economic Region (SER) (1982), Suzhou-Wuxi-Changzhou Region (1985), and Bohai Sea Rim (1986) (Zhang et al. 1987). These centrally mandated but loosely organized and informal forms of government coalitions and alliances were expected to serve as a key institutional pillar to break out of the hierarchically organized production and planning in order to create the impetus for territorially integrated growth.

By the mid-1990s, economic regions established in the 1980s had been widely discredited because there had been little progress. Many of them were dismantled. The SER, for example, was disassembled in 1987, five years after its inception (Wu 2006). To accompany the abolition of many economic regions, a new mosaic of regional development and territorial competition began to crystallize in China. The economic infrastructure of major city regions has been significantly restructured in conjunction with: the acceleration of globalization and market reform; the expansion of city-based accumulation strategies (e.g. urban space commodification by land and housing reform); the deterioration of decentralization-induced fragmentation; and the rise of urban entrepreneurialism (Xu 2008). Under these conditions, local governments begin to adopt new competitive strategies in order to adjust to reform-imposed fiscal constraints and to attract external capital, and "localism" becomes a severe problem. Using the concept of "economy of administrative division" (EAD) (*xingzheng qu jingji*), Liu (2001) defines Chinese localism as a special feature of economic organization in China, in which market activities and economic development are normally organized on the basis of individual administrative units. It is argued that EAD—the visible wall—exacerbates pro-competition culture. The resultant inter- or intra-regional "resource wars," "production wars," and the proliferation of internal trade barriers and market separation highlight the fragmentation of China's economy, in which different jurisdictions are like rival "economic dukedoms." With this rapidly developing political environment, the consequences of administrative fragmentation—such as increasing subsidies and giveaways to investors, inefficient and uncoordinated duplication and oversupply of infrastructure, facilities, and services, overriding diversion of scare public resources away from environmental and social concerns to economic growth—become more and more acute (Xu and Yeh 2005).

Since 2000, the problems of political fragmentation within mega-city regions have resulted in many new programs of regional governance. Informal organizations for city networking have been set up to provide a framework for political negotiations regarding a number of regional issues. In some regions, such as the PRD, Yangtze River Delta, Beihai Bay, Northeast Six Coastal City, and Taiwan Strait West, new regional planning efforts have been made, attempting to define a coordinated and sustainable regional growth. These newly formed organizations for city networking are no longer being promoted only in response to internally triggered demands and imposed or mandated from above; they are emerging as a product of bottom-up initiatives and interplay of internal and external forces as the process of globalization and market reform unfolds (Xu 2008).

To explain what is behind the renewed interest in mega-city regions, Xu (2008) summarizes three reasons why the mega-city region matters in China today. First, for central and provincial governments, political fragmentation due to decentralization has increasingly weakened their governing capacity. This has given rise to an urgent need to recompose state strategies for better regulation. Strategic planning at the mega-city region level has thus become a key political strategy to regain control and reassert the functional importance of provincial and central governments in local and regional economic governance. Second, for provinces, planning the mega-city regions can help strategically reposition their "regions" in the national and global economic landscape. Third, provinces also rely on creating mega-city regions to obtain central policy support. Xu and Yeh argue, in Chapter 1 of this volume, that decentralization does not necessarily mean the decline of the central state in post-reform era, because the central state is still an important level of strategic intervention although its dominance is now being decentralized. The approach is often "top-down," involving the central attempts at strategic distribution of economic activities.

One major focus is on the location of large manufacturing projects. Policy tools include some direct measures, such as location control of firms and the movements of state-controlled industries and services. More indirect measures seek to increase the attraction of intended areas for inward investment by providing major infrastructure and preferential policies. Being an iconic region incorporated into national development plans (e.g. the National Five-Year Plan) is an important indication of the central government's support and possible resource allocation (Xu and Yeh 2009a). Indeed, creating city regions is believed to be able to attract central attention, because the PRD, Yangtze River Delta, and Binhai New Region in Tianjin set good examples for others regions to follow. As inter-regional competition for national status is escalating, provinces believe that new regional spaces can help them stand out strategically.

Consequently, one striking characteristic of current mega-city region planning is its explicitly controlling and competitive orientation. Proposals have been justified as a means of regaining government control over fragmented growth and repositioning regions in the national and global economic landscapes. In contrast to the 1980s, when debates on regional cooperation simply focused on the issues of administrative coordination and city networking, current discussions have also been oriented towards the overarching priorities of policy consolidation and the promotion of the structural competitiveness of regions (Xu 2008).

Historical geography of the Pearl River Delta: a growing polycentric region

The PRD is one of the most frequently researched mega-city regions in China. It covers a land area of 54,744 square kilometers with a resident population

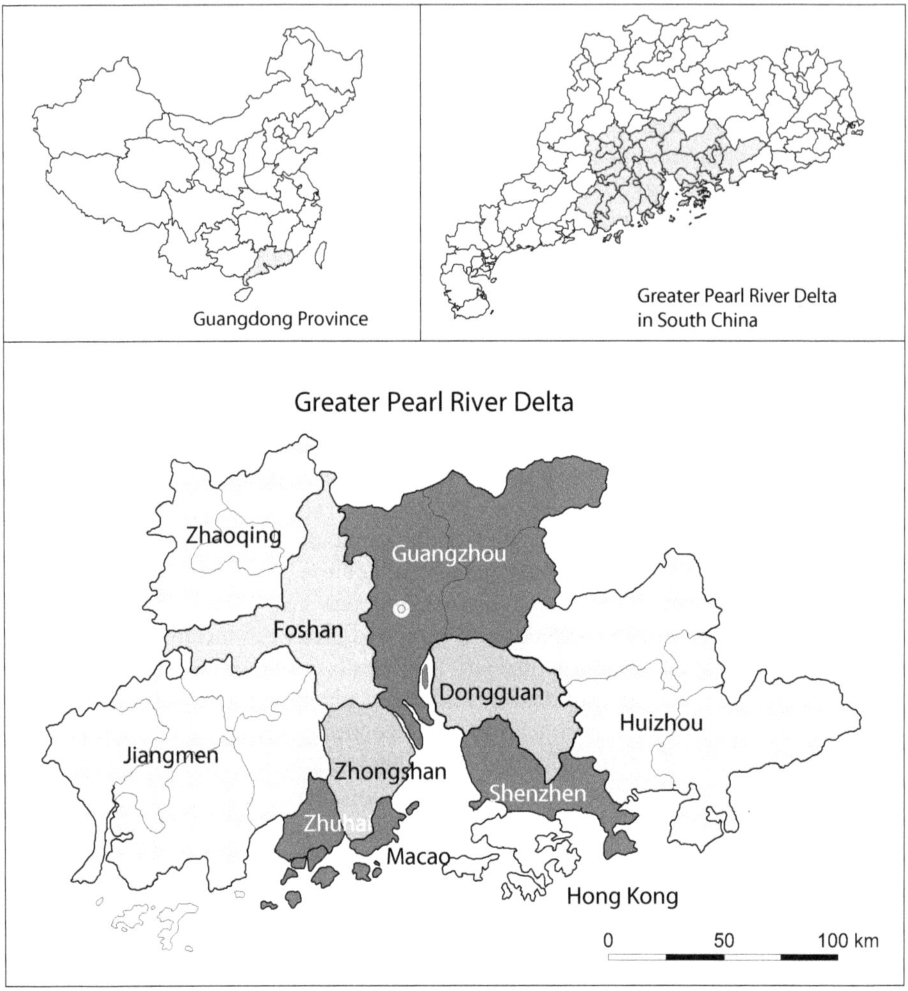

Figure 10.1 Location and component cities of the Pearl River Delta
Source: Chapter authors.

of about 50 million (floating population included). It comprises nine municipalities (Figure 10.1). Regional transformation in the delta has been among the most dramatic in China. Before 1978, the delta was largely an agricultural region with only one major clearly defined city in Guangzhou. As an ancient civilized city, Guangzhou has a long history of more than 2,100 years, the same as the world-renowned ancient city of Rome. Three legendary nicknames of Guangzhou—the Flower City, the Goat City, and the Rice-ear City—testify to its ancient origins. Situated in southern China with a number of hills in the northeast and the PRD alluvial plain in front, Guangzhou lies at the confluence of the Dong (East) and Bei (North) rivers and has a very advantageous geographic location (Xu and Yeh 2003). Because of its location at the northern tip of the PRD, which is fed by more than 1,000 miles of waterways, Guangzhou possesses an exceptional port, and has been nicknamed the "Silk Road on Water" (Xu 1985). About ninety-five years ago, two train lines connected Guangzhou with Hong Kong (through Dongguan and Shenzhen) and Shanshui (now within the jurisdiction of Foshan) respectively, adding to the development of Guangzhou as the largest city in the PRD and a major export port in southern China. However, the change of regime in China in 1949 modified the trajectory of Guangzhou. At that time, China was entirely closed to the outside world under a self-reliance policy. The importance of Guangzhou thus declined while Hong Kong surpassed Guangzhou to become the trading extrepôt and a booming industrial city.

Since 1978, economic reform has offered unprecedented opportunities for Guangzhou and other delta cities. During the late 1970s, Hong Kong experienced difficulties due to rising production costs and the emergence of cheaper manufacturing centers in the Association of Southeast Asian Nations (ASEAN) countries (Sit 2001). Fortunately, such pressure coincided with the opening-up of Mainland China. Guangdong Province, which is next to Hong Kong, was designated as the vanguard for implementing new open policies. In particular, the PRD was selected as an open economic zone in 1984, with a number of granted one-step-ahead policies to attract overseas capital.

Prior to 1984, the delta did not have a strong internal nor external identity as a region, although component cities are mostly contiguous and located in the Pearl River Basin. The delta lacked any form of governance and planning arrangements, and there was an absence of a shared sense of belonging. In terms of economic functioning, most parts of the delta barely appeared interdependent. So, the idea of artificially designating the delta as a region mainly focused on the steadfast commitment of Guangdong to attract overseas capital. Intercity networking was not a priority at that time.

This strategy to attract capital proved an immense success. First, starting in the 1980s, a number of Hong Kong industrial capitalists exploited this unexpected and hitherto untapped cheap labor and land in the PRD to establish branch plants or undertake sub-contracting processes there. This was followed by a massive relocation of manufacturing activities from Hong Kong to the PRD a decade later. Hong Kong was soon restructured to a regional service center, providing its industrialists in the PRD with "front-end" functions such as research, marketing, and distribution, while the PRD took advantage of capital flow to develop a clustering of factories and a revived international workshop. A regional division of labor of the "front shops, back factories" model began to take shape. To support rapid economic expansion in the PRD, delta governments and overseas investors have invested heavily to improve transportation and other infrastructure development (e.g. the Guangzhou-Shenzhen Highway), reinforcing the delta's internal and external identity as a region.

In more recent years, this synergy created by the regional division of labor has gained further momentum as the delta has developed very rapidly to include a wide range of economic activities. For example, the delta has specialized as a world leader in the production of a range of technological products, especially in the realm of advanced electronics. It is also developing a plethora of extremely competitive clusters for the production of light manufactured and electronics goods. There is a strong local or locational integration of numerous complementary firms that produce the inputs and components to support its world-leading industries of toys, footwear, and electronics (Enright et al. 2005). At the same time, many delta cities are now making efforts to reposition their roles in the global and domestic market through placing large investments in heavy industry, as exemplified by the strategic decisions to develop the motor industry in Guangzhou and petrochemical plants in Huizhou and Zhuhai.

With such a huge transformation, the PRD has become one of the regions with the highest degree of fragmentation in China. Several forces reinforce this trend. First, the proliferation of overseas capital-driven economic activities brought forth a laissez faire culture in the delta from the very beginning. Cities, towns, and villages have had to mobilize their own resources through personal linkage, ancestral networks, and other local initiatives to partner overseas investors (Smart and Smart 1991; Lin 1997; Sit and Yang 1997; Hsing 1998). This distinctive decentralized form combined with proper incentives allows local governments to play an increasingly corporatist role (Oi 1995; Walder 1995). Decentralization soon becomes a self-reinforcing process—more and more economic power is further devolved to cities, towns, and villages.

The second force is decentralized land development. Normally, the state monopolizes the primary land market, and collectives are forbidden to sell or lease land to developers. These limitations, however, never constrain the spread of the black market, where primitive and very often ungovernable land transactions are so pervasive (Wu et al. 2007; Xu et al. 2009). A new land policy has been implemented in Guangdong to permit collectives to trade rural construction land with developers. This policy has injected renewed incentives for bottom-up initiatives in land development.

Third, a dispersed governance structure leads to an even more powerful force of fragmentation. The PRD has one of the most sophisticated jurisdictional structures, which includes the provincial capital, Guangzhou, and two special economic zones, Shenzhen and Zhuhai. Guangzhou and Shenzhen are two sub-provincial-level municipalities that are given a level of power higher than a prefecture but still lower than that of a province. Such special status means a high degree of autonomy, and they are thus relatively outside the control of the provincial planning system. These sub-provincial municipalities are followed by seven prefecture-level municipalities. Below prefecture level are country-level divisions, which include eight county-level cities, two counties, and thirty-one urban districts. This is followed by more than 230 towns. Jurisdictions at these different levels have idiosyncratic status and enjoy specific power in urban development and planning.

Administrative fragmentation has a great spatial implication. In the past, the PRD was focused mainly on one clearly defined central city (i.e. Guangzhou); however, the delta has now become increasingly polycentric. Some former peripheries or rural areas have developed into important industrial cities, challenging the traditional center of production and service in hosting economic activities. Guangzhou is still a strong economic center and continues to play its strategic role. For instance, it remains as the place where the majority of the tertiary sector is developed. Guangzhou's role as a regional distribution center is represented by its dominating functions in sectors such as transportation, port operation, and external trade. However, the dominance of Guangzhou is now challenged by the rise of other delta cities, such as Shenzhen, Dongguan, Zhongshan, and Foshan. Thirty years ago, Shenzhen was a tiny fishing village with a population of 30,000. It was established as a special economic zone in the early 1980s by the central government due to its proximity to Hong Kong, and has thrived ever since. Nowadays, the city hosts more than 12 million people (including unregistered floating migrant workers), and surpasses Guangzhou in a number of economic indicators such as GDP per capita and external trade. Shenzhen is a regional center of shipping and logistics. Dongguan, Foshan, Zhongshan, and Huizhou have quickly transformed into strong industrial cities in their own right.

Using location quotient (LQ) values, Weng (2006) performed an analysis to identify the local specialization and functional distribution of industries in the PRD. It was found that each municipality has own specialized industrial sector. Municipalities on the eastern side (Guangzhou, Dongguan, Shenzhen, and Huizhou) present their strengths in electronics and telecommunication equipment, instruments, meters, cultural and office machinery, electric power, and heat power, whereas those on the western side (Foshan, Jiangmen, Zhongshan, Zhaoqing, and Zhuhai) have put more focus on the textile-related industries and metal products, and electrical machinery and equipment (Weng 2006). Generally speaking, most municipalities have specialized in similar industries. This suggests the existence of possible competition in these industrial sectors among the nine municipalities. Of the government-defined leading industries in these cities, almost all of them have a focus on industries such as electronics and information, biotechnology, new materials, and new energy.

Weng's study of economic structure suggests a high degree of economic fragmentation—the nine delta cities are like discrete entities that seem to be barely complementary to each other in their economic roles. Indeed, fragmentation also appears in areas such as infrastructure planning. Many scholars argue that there are probably too many seaports and airports in the delta. In the early 1992, the central government approved the construction of three deep-water ports adjacent to Hong Kong—namely, Yantian in Shenzhen, Huizhou port in Daya Bay, and Gaolan port in Zhuhai. These are in addition to the existing Guangzhou port and Shekou port in Shenzhen. There will be inadequate freight for all these ports (Yeh 2001). Similarly, there is also duplication in airport facilities. Within the vicinity of Hong Kong, in a small radius of 25 kilometers, there are four big airports—Hong Kong, Shenzhen, Macau, and Zhuhai. Although it is technically debatable whether or not these facilities are really too many for an area with a population of 50 million, each government looks after its own interests and desires a seaport or an airport as a showpiece. Cities build their own facilities such as reservoirs, sewage treatment plants, and roads. There is little city-networking in regional service provision.

PRD cities are also plagued by their collective environmental externalities. Rapid population and industrial growth has led to a sharp decline in overall environmental quality. For instance, water pollution in the municipal and agricultural sectors is increasing because large amounts of municipal wastewater are being discharged into surface water bodies with minimal or no treatment, contributing to an increasing load of organic and toxic waste in major waterways (Lee 2001). Conflicts over major waterways (e.g. Dongjiang or East River) are most profound between upstream and downstream cities. Upstream cities are forcefully persuaded to forego economic development projects in the name of conservation and protection of headwater and are increasingly vocal in demanding just compensation from downstream ones who have prospered through rapid industrialization and whose industrial effluent has increasingly polluted the Dongjiang from the mid-stream section onward (Lee 2006).

Space of flows: interaction among municipalities in the Pearl River Delta

The aforementioned studies on the history and economic structure of the PRD delineate how the delta emerges as an entity with a high degree of fragmentation. To what extent can we assess the PRD as an integrated region? What is the current status of intercity linkage? Studies on the "space of flows" in the delta are still rare. However, the China Academy of Urban Planning and Design (CAUPD) made an initial effort to examine the interaction among delta cities by obtaining and analyzing a dataset on daily, city-to-city, two-way vehicle volume. CAUPD then mapped vehicle flows for major highways (including express and national highways) in the delta (see Figure 10.2). On this schematic map, the width of a line indicates the daily average two-way vehicle flow of a major express highway or a national highway. Although vehicle flows do not tell the exact nature of interaction between pairs of cities (e.g. goods movement traffic), they suggest the relative strong or weak linkage among PRD locations. For instance, the map clearly shows the central role of Guangzhou as a transportation hub and the stronger linkage on the eastern side (Guangzhou-Dongguan-Shenzhen). On the western side, the interaction is only relatively strong between Guangzhou and Foshan, while other cities show weak signs of connection.

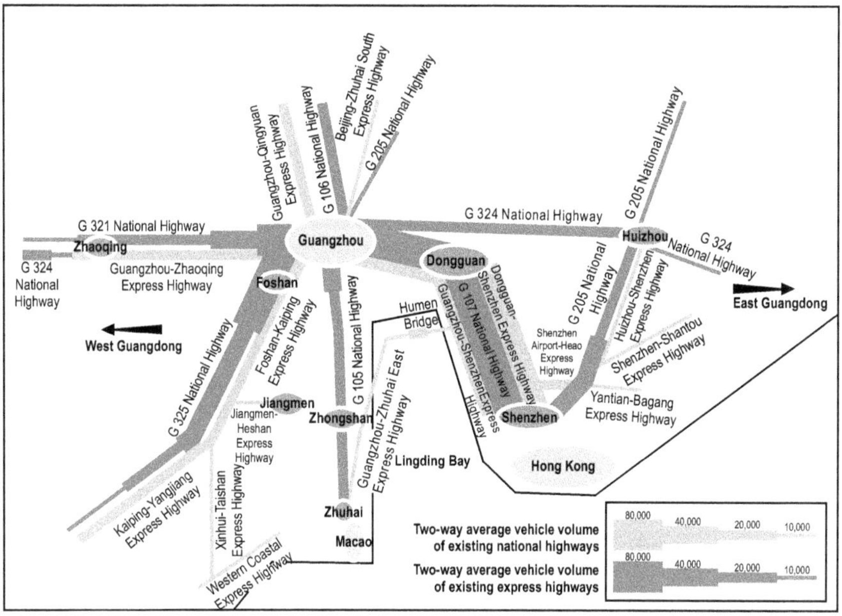

Figure 10.2 Major highways vehicle volume distribution

Source: Adapted from China Academy of Urban Planning and Design 2004.

In addition to vehicular flow, commuter bus, train, and ferry services provide additional dimensions to assess the transportation linkages between cities (Xu and Yeh 2009b). Our previous study identifies several features of such linkages:

1. The commuter bus and train services suggest the central role of Guangzhou as a transportation hub. Shenzhen and Hong Kong are also major nodes. In particular, Shenzhen, because of its location between two large cities (i.e. Hong Kong and Guangzhou) and the presence of a political border (a major factor that causes the decline of interaction between Hong Kong and delta cities), may have captured much flow (e.g. cargo, vehicle, and pedestrian) because of intervening opportunity.
2. There is a strong linkage between Foshan, Guangzhou, Dongguan, Shenzhen, and Hong Kong, while connectivity among other cities is relatively weak.
3. Although the western side is fed by more than a thousand miles of waterways, ferry commuter services are barely developed. Railway transport is not available. Bus services are relatively well organized among locations, but these involve higher transportation costs (e.g. toll charges).

One major conclusion drawn from this analysis is that many delta cities (e.g. Foshan, Guangzhou, Dongguan, and Shenzhen) may be more closely connected in performing their functions than has commonly been believed. Weng's LQ-based analysis discussed in the previous section was conducted on the basis of only a sketchy industrial classification. For instance, it may reveal that both Dongguan and Shenzhen are engaged in the manufacturing of electronics products; it cannot provide precise and detailed information on each city's specialization in the sector. Enright et al. (2005) undertook a more detailed analysis to identify industrial specialization in the delta. It found that although delta cities may compete in some sectors, a wide range of industrial sectors and/or sub-sectors have a strong geographic concentration in individual cities or towns. For instance, while Dongguan specializes in computer parts, Shenzhen focuses more on computer production. This helps to develop certain types of complementary economic functions among locations, which may explain the strong tie among cities in the eastern side.

Such analysis suggests the potential to treat the eastern side as an interconnected region, but it is not very convincing about treating the PRD as a single region because the western side does not have strong ties among its cities and with its eastern counterpart. This demands a strategic planning consideration if the delta is to be planned as an entirety.

Planning for the fragmented Pearl River Delta

Since the late 1990s, the PRD, like many other Chinese regions, is under increasing pressure to take a more proactive stance if it is to prosper. Inter-region competition for central policy support and resources is intensifying. The State Council approves the plans of the Tianjin Binhai New Area and the Chengdu-Chongqing Region as free trade zones and experimental fields for further economic reform. Together with Pudong in Shanghai, they are counted as three new economic engines. This trend works to the disadvantage of Guangdong, which wants to keep the PRD as a national economic hub. Guangdong believes that regional planning can make the delta more competitive. The province suggests that the PRD will be developed into a renowned international manufacturing workshop and global city region by 2020. By then, the population is expected to reach 65 million, but infrastructure is to be designed to allow for a population of 80 million. How can the region prepare to accommodate the expected growth and development objectives of the province? In this section, we explore the planning institution for mega-city regions in China and then use the PRD Urban Cluster Plan as an example to answer this question.

Planning institution in mega-city regions

Chinese government is hierarchically organized at the central, provincial, prefecture, county, township, and village level. There is no formal government structure at the city region level. National laws do not require cities to cooperate with adjacent jurisdictions, nor does the central government offer any incentives to forge such cooperation. If a city region (e.g. PRD) contains jurisdictions under a single province, the provincial government can be a key facilitator in coordinating regional development, but if the city region (e.g. Yangtze River Delta) extends to include jurisdictions under different provinces, the central government is essential in coordination. Theoretically, the central government and provinces can build on China's strong state tradition to enforce coordination. However, the institutional structure for state intervention is problematic, because the functions of regional planning are highly fragmented among different ministries (Xu 2008).

The National Development and Reform Commission (NDRC) is the most important organ in the central government. It is not only the agency in charge of devising major economic development strategies of the country by making the five-year socio-economic plans but it is also a powerful actor in authorizing state resource allocation and major projects. Socio-economic plans have been in operation since the pre-reform period, and have recently come to contain a strong spatial element. The NDRC is a central agency that is a half-level higher than other ministries for historical reasons. Therefore socio-economic planning has the capacity to guide, coordinate, and constrain spatial plans made by other ministries. Socio-economic plans are made to clarify the overall positioning of regions and major cities, provide blueprints for priority development areas, and find solutions for problems that are difficult for one city or one province to solve (Hu 2006). More importantly, socio-economic planning directly connects state resource allocation to spatial formation.

The Ministry of Housing and Urban-Rural Development (MHURD) and its local subordinates such as construction commissions and municipal urban planning departments are organizations traditionally responsible for the physical or spatial planning of cities. To achieve better planning control, they also prepare two kinds of regional plans. One is the metropolitan plan, which is used to coordinate development of central cities with their outlying countries and towns. The other is the urban system plan, made to rationalize the functional, hierarchical, and spatial structures of infrastructure and urban settlements in a multi-centric region. Both plans are said to provide necessary elements of spatial coordination, such as linkages of cities, distribution of regional infrastructure facilities, and other spatial attributes such as industrial space, transportation hubs, wilderness, and conservation areas. These plans are normally incorporated into socio-economic plans made by the NDRC as supplementary sector guidance. Nonetheless, in recent years, the rapidly changing functional and territorial contours of regions urgently require spatial plans to have a higher status. Consequently, some provinces invite the MHURD to make spatial plans for their regions, bypassing the NDRC. These plans pay particular attention to population and region-wide environmental, social, and economic issues, and develop an extremely strong element of strategic consideration, which is similar to the function of the NDRC's socio-economic plans. However, the MHURD's capacity is punctuated by the lack of concrete measures to link the planned "spatiality" with the resource allocation of national economic planning, the enforcement of development control, or any tangible socio-economic policies (Wu et al. 2007).

The Ministry of Land and Resources (MLR) and local land departments are entitled to prepare land use plans at all levels, but primarily address issues of land conservation and farmland protection. In more recent years, new contents have been added to these plans to provide land use projection for development projects and to demarcate different zones for regulation—an element contained within physical or spatial planning (Xu 2008). Last, but not least, other ministries such as the Ministry of Environmental Protection (MEP), the Ministry of Railways, the Ministry of Transportation, and the Ministry of Agriculture have their own regional plans to guide sectoral development.

The fragmented functions of regional planning attribute to inter-ministerial conflicts. The NDRC used to focus on project-based resource allocation and did not pay much attention to planning for regions, even though it is accorded with the power to provide general guidance to socio-economic growth. But recently, the NDRC has realized the potential and importance of comprehensive planning at mega-city region level, and dislikes the idea of two rival departments competing for the regional planning market. Likewise, the MHURD claims that its plans have spatial elements and should be taken more seriously, because the other two departments do not have adequate expertise to make spatial plans. The MLR's argument is based on the fact that its plan has a wider geographical coverage and belongs to upper level spatial guidance. But more basically, the planning structure has fundamental difficulties—the NDRC, MHURD, and MLR have roles that overlap. One result is a demarcation dispute among them. With such a flawed institutional structure, regional governance and planning may be subjected to strong influence by politics and may have weakened the power of central and provincial governments in regional coordination.

Strategic planning in the Pearl River Delta

Since the late 1980s, eight regional strategic plans have been produced for the PRD. They are formulated by the Guangdong government in collaboration with four different ministries, namely, the NDRC, MLR, MEP, and MHURD. The NDRC made three regional plans (i.e. socio-economic plans)—in 1994, 1998, and 2009. The first two plans tried to coordinate growth in the delta through state resource allocation, but they failed to do so because market forces became increasingly beyond state control. The 2009 plan is a broad-brush guidance and does not offer concrete measures for spatial coordination. The MLR made a regional plan in 2005 primarily to address the issue of farmland protection, while the MEP's 2003 plan was focused on environmental issues.

The MHURD made two PRD plans—in 1990 and 1994—to offer planning guidance on functional and spatial distribution of the delta cities. The 1990 plan failed because it was intended to manipulate regional spatial development such as city size superficially, which is difficult to control even under a centrally planned system (Wu et al. 2007). The 1994 plan was made as a supplementary document of the NDRC's 1994 plan. The MHURD's 1994 plan—a highly praised, award-winning plan—is said to have failed (Yang 2004; Zou 2006). In 2004, Guangdong invited the MHURD to make another regional plan which is commonly called the PRD Urban Cluster Coordinated Development Plan (UCCDP). This plan was designed to set out new spatial regulations to achieve the failed goals of the 1994 plan. An UCCDP Ordinance was also promulgated in 2006 to enforce implementation of the plan. UCCDP is considered the most comprehensive and skillfully made plan among all plans for the delta.

Table 10.1 summarizes the strategies proposed by UCCDP. These strategies tackle issues in relation to the development objectives, population, land use, spatial policies, infrastructure, environment, and major actions plans. A growing population is expected in the delta, and this will be associated with upward pressure on increased sprawling land use, inadequate infrastructure, and declining environment and natural habitats. UCCDP proposes to address these challenges through rapid economic growth, integrated provision of infrastructure, and better top-down regulation. The plan accentuates policies on integrated service provision and transportation. It provides guidance on how to provide an integrated and balanced transport system, stressing the need to maintain and improve transportation networks, and to facilitate intercity transit systems. The economic justifications of this transport system have been closely intertwined with escalating inter-region competition.

One important policy initiative of UCCDP is the use of a hierarchical approach in spatial planning. To this end, the delta is divided into nine policy zones that are placed under four different levels of spatial control, namely supervisory regulation, directive regulation, coordinative regulation, and conductive regulation. Areas under supervisory regulation are subject to the strictest top-down control. These include open space and transport corridors. Cities and towns in these areas are not allowed to make independent investment decisions like they used to, nor can they alter defined uses. Provincial government is the primary decision-maker. Accordingly, areas under conductive regulation are under the least top-town control. Practically, such a planning approach has the potential to divert development away from areas under protection. Politically and perhaps more importantly, it amplifies the supervisory and regulatory functions of the provincial government as a focal point for economic governance and spatial coordination. It is exactly these efforts that evidence a novel trend of state reconstruction in which the state is rebuilding its erstwhile power and its institutional capabilities, legitimacy, and territorial mapping.

Table 10.1 Strategies recommended in UCCDP

Development objectives and scale	Leader in global cooperation; National economic engine
	Cultural demonstration region; Reform experimental region
	Pioneer in regional coordination and urban-rural integration
	Population: 60 million by 2050
	Land use: Control urban land use
	Conservation of eco-regions
Spatial development strategies	Enhance the role of central cities and develop growth spines
	Increase cooperation with inland provinces and waterfront development
	Promote growth in the West and improve development in the East
	Emphasize territorially balanced development
	Introduce eco-regions
	Improve living environment through completing infrastructure provision
Spatial layout strategies	Identify and define functions for development corridors, nodes, and spines
	Introduce a hierarchical system of central cities
	Define functional distribution of cities
	Identify and define different kinds of clusters and character areas
	Provide guidance for three major metropolitan areas
Regional supporting system	*Environment*
	Identify locations for eco-conservation
	Define conservation policies for each eco-region
	Protect critical habitats
	Invest in green infrastructure networks of open space
	Strict control of development in eco-regions
	Use greenbelts to control urban expansion
	Protect cultivated land
	Protect water resources
	Address water supply, drainage, and point pollutants management
	Encourage effective energy use and reduce air pollution

Table 10.1 Continued

	Encourage coordinative environmental infrastructure planning
	Identity
	Preserve cultural landscape and heritage regions
	improve community services
	Transportation
	Improve external connection
	Improve transportation efficiency
	Enhance intercity links
	Emphasize intermodal connections
	Complete transportation network
	Establish regional integration in transportation
Policy zones and spatial regulation	Define nine policy zones
	Establish policy guidance for each policy zone
	Set up four levels of spatial regulations
Spatial coordination planning	Conserve land in new projects
	Promote compact land use
	Develop affordable commodity housing and social housing program
	Consider environmental sustainability when locating new development areas
	Continue bio-diversity
	Address rural development
	Provide respective spatial coordination guidance for nine PRD cities
Major action plan	Enhance external connection
	Develop bay areas
	Implement green-line policy to protect eco-conservation areas
	Promote heavy industry
	Materialize regional transportation integration
	Introduce waterfront development
	Address small-town management
	Establish a platform for data-sharing
	Establish a provincial office for coordinating planning issues
	Enact an ordinance to enforce plan implementation

Source: China Academy of Urban Planning and Design 2004.

Rising concerns and the politics of planning

Regional coordination through UCCDP has undoubtedly bred tension, particularly with a hyper-competitive environment tending to prevail and with local interests being undermined to various degrees. This plan is not simply a policy articulation. More deeply, it is a political representation of how different cities are represented. This issue of representation is an important part of the persuasive capacity of strategic planning. However, as Xu (2008) pointed out, UCCDP has led to dissatisfaction among many cities because it has a "dangerous obsession" with conveniently safeguarding powerful cities at the expense of others. This is particularly evident in the way that the plan proposes a concentration of large capital projects in a handful of cities (e.g. Guangzhou), while other cities are being peripheralized. Zhuhai argues that it is unfair to treat its city only as a sub-center in the delta, while Guangzhou and Shenzhen are two prime centers. This sub-center status places Zhuhai in a disadvantageous position for obtaining central and provincial support to locate large projects. In one case, Kuwait Petroleum Corporation originally planned to invest USD 5 billion to build an oil refinery in Zhuhai. However, it later decided to locate the project in the Nansha District of Guangzhou to follow the requirement of the provincial government (Xu 2008).

As the capital of Guangdong, Guangzhou is more or less privileged in UCCDP because the city cadres hold considerable sway in the plan-making process, contending with planners over its interests. The city is finally being allowed to build a heavy industrial zone in Nansha, an ecologically sensitive location in the center of the delta, to accommodate investors from automobile, steel, mechanical equipment, electronics and petrochemical products industries. This situation was permitted because the planners were pressured to deliberately classify Nansha as an area under conductive governance, and this is subject to the least top-down supervision and planning control. Guangdong believes that placing the development focus on Guangzhou, the provincial capital, can increase economic efficiency to better reposition the delta (Zou 2006). Partially because of the unbalanced political representation, the implementation base of UCCDP is weak. Cities refuse to incorporate defined spatial policies into their own municipal plans. One consequence is that UCCDP pays only lip service to uncertainty. Admittedly, the plan does give indications as to the actions that would lead to the achievement of the desired objectives, but it leaves it more or less at that, due to inadequacy in persuasive capacity.

Other concerns arise because UCCDP is drawn up by the MHURD, which is a sectoral department and does not have the authorized capacity to implement a comprehensive regional plan involving multi-sector and inter-scalar coordination. Moreover, strategies depicted in UCCDP may not be fully supported by other ministries such as the NDRC and MLR, who control state resources (i.e. capital and land) that are critical in plan implementation. The MEP, for example, has recently overruled the decision to locate the Kuwait Oil Refinery in Nansha. Worse still, the State Council does not approve UCCDP, because it holds the opinion that regional plans should be made by the NDRC, who is the authorized agency that coordinates with different ministries and their sectoral plans.[1] This works to the disadvantage of Guangdong in obtaining central support and repositioning itself in the national economic landscape. In 2009, Guangdong, under the leadership of a new Party Secretary, made tremendous efforts to lobby the NDRC to make another socioeconomic outline plan for the PRD. The province succeeded in persuading the NDRC to make the plan, largely due to the new Party Secretary being an important member of the Politburo (the Political Bureau of the Communist Party of China's Central Committee, the highest decision-making body in China). Afterwards, the 2009 plan was approved by the State Council, and thus carries great significance and influence because it is seen to fully reflect central policy guidance on the delta's development and signify a "green light" by the central state to implement further experimental reform in the delta. However, the NRDC's 2009 plan takes an overview approach and does not articulate policies into concrete spatial strategies. One overriding concern is the lack of attention to important regional issues such as cross-boundary coordination, social problems, and sensible spatial strategies of different cities. The plan's failure to highlight institutional reform to address inter-scalar tension and urban coordination is also viewed as a major oversight.

Concluding comments: towards a state-theoretical interpretation of regional strategic planning

The theme of this chapter is China's regional strategic planning, focusing on the changes in spatial formation and planning responses in the PRD. We briefly review the origin and growth of the delta, and examine spatial fragmentation and functional interactions among nine delta municipalities. Prior to 1984, the delta was largely an agriculture area and lacked internal and external identity as a single region. It was artificially established in 1984 primarily to attract overseas capital. After more than twenty years of development, the PRD has grown rapidly into a renowned international manufacturing zone. Spatially, it exhibits an increasingly polycentric pattern; functionally, it demonstrates a certain degree of linkages. The mismatch between the fragmented administrative boundaries and the functional–economic territory in the PRD requires a strategic vision to plan the region in its entirety. It is in this context that regional strategic planning is now increasingly being mobilized as a mechanism of economic development policy and a political device through which the state is attempting to enhance place-specific socio-economic assets and to regain control in a growing sophistication of territorial development. Indeed, the PRD's strategic planning is only one instance of a wider response to the rise of "regional renaissance." From the Northeast Six City Coalition, the Yangtze River Delta, the Beijing-Tianjian-Hebei, and Taiwan Strait West to Beihai Bay, Chongqing-Chengdu, and the Pan-PRD, local and regional economic policies are becoming increasingly linked to diverse forms of spatial planning, infrastructure investment, and political-economic coordination at the mega-city region level. It is in considering this very context that we unbundle the "currently unfolding regional regime" in China. In doing so, we develop a new state-theoretical interpretation towards understanding regional strategic planning under transition.

The majority of existing studies on urban and regional transformation in China have focused on one particular area of urban and regional transformation, that is, economic reform and the ensuing rise of localities and the market as key forces in socio-economic development. There are many theoretical discourses that have been developed to capture this critical moment and to demonstrate how such profound reshufflings has led to a powerful trend of fragmentation and decentralization in the economy. A decentralized and de-hierarchized structure has been justified as a basic institutional precondition for promoting territorial growth. Admittedly, decentralization and fragmentation have grown remarkably in the past decade, but even this trend is overshadowed by a wave of consolidation that cannot be ignored. The more the structure is fragmented, the more centralized the top-level control operation might become. All things considered, this chapter argues that a regulatory strategic planning that accentuates the hierarchical linkage for a more centralized power is becoming the new way for the state to get involved in space commodification. Regional strategic planning presents only one dimension of changes in which counter-trends might be constructed. Recentralization happens in many other areas and on different scales. In part, then, this chapter joins an emerging rank of contribution (Xu and Yeh 2005, 2009a; Xu 2008; Yeh and Xu forthcoming) that question the firmly believed notion that "urban and regional transformation in China is a major consequence of the decentralization process." On the contrary, we need to rethink our theoretical approach to the central and provincial state so as to unravel their persistent roles as key sites, functions, and orchestrators of territorial growth. In this sense, new interpretations of urban and regional transformation in China should be understood by underscoring the interplay between trends of decentralization and counter-trends of recentralization (Xu and Yeh 2009a).

The extent to which the political regime will succeed in challenging any potential erosion of control and rights in its planning efforts to retain some degree of regulation and cohesion would provoke some interesting debates. This is due to the historical contingencies and path-dependencies in a transitional society. Institutionally, China's regional planning system carries strong legacies of the former state socialism. There is a lack of structure to forge horizontal networking because China does not have a tradition of valuing this type of urban partnership, and the dominant linkages are hierarchically organized. In this sense, regional growth has to depend on strategic intervention of supra-units, either provincially or centrally. With a long established tradition of a strong state, the hierarchically organized state may have the potential to develop a good capacity for government-led strategic interventions at the regional level, but the function of regional planning is fragmented among different ministries. This results in inter-ministerial rivalries and real difficulties in coordination, and thus weakens the top-down governing capacity. Regional strategic planning may appear to be little more than a cosmetic makeover that hides the intensifying inter-scalar and inter-territorial competition within major city regions in China.

Acknowledgment

The research is supported by the General Research Fund of the Research Grants Council of Hong Kong (Project code: CUHK453609).

Note

1 Information was obtained from several interviews conducted in Beijing in June–July 2009 with government officials and planners of the National Development and Reform Commission and Ministry of Housing and Urban-Rural Development.

References

China Academy of Urban Planning and Design (2004) *Pearl River Delta Urban Cluster Plan*, Beijing: China Academy of Urban Planning and Design (in Chinese).

China City Mayor Association (2004) *Chinese City Development Report*, Beijing: Commercial Printing House (in Chinese).

Enright, M., Edith, E.S., and Chang, K.M. (2005) *Regional Powerhouse: The Greater Pearl River Delta and the Rise of China*, Singapore: John Wiley & Sons (Asia) Pte. Ltd.

Hsing, Y.T. (1998) *Making Capitalism in China: The Taiwan Connection*, Oxford: Oxford University Press.

Hu, X.W. (2006) "Evaluation and Prospect of China's Regional Planning," *City Planning Review*, 30(supplement): 8–12, 50 (in Chinese).

Lee, Y.S. (2001) "Environmental Issues Arising from Rural Industrialization and Decentralized Urbanization in the Pearl River Delta," in Greenpeace China (ed.), *China Environment Report: Ecological Lessons of the Pearl River Delta*, Hong Kong: Greenpeace China, pp. 23–35 (in Chinese).

—— (2006) "Managing Water Resources in the Delta Border Zone: Challenges and Opportunities," in A.G.O Yeh, V. Sit, G. Chen, and Y. Zhou (eds.) *Developing a Competitive Pearl River Delta Region in South China*, Hong Kong: Hong Kong University Press, pp. 357–81.

Li, Z. (2003) "Facilitate Urbanization to Develop the Central Role of Cities," *Management Review*, 15(1): 9–11 (in Chinese).

Lin, G.C.S. (1997) *Red Capitalism in South China: Growth and Development of the Pearl River Delta*, Vancouver: UBC Press.

Liu, J. (2001) "On Administrative Organization and Innovation of Management Model in Metropolitan Region in Mainland China: Reform on Administrative Setting in the Pearl River Delta," *Economic Geography*, 21(2): 201–12 (in Chinese).

Oi, J.C. (1995) "The Role of the Local State in China's Transitional Economy," *The China Quarterly*, 144: 1132–49.
People's Daily (2002) "Experts Say China's Urbanization Rate Tipped to Rise," *People's Daily*, July 20, 2002.
Sit, V. (2001) "Economic Integration of Guangdong Province and Hong Kong: Implications for China's Opening and Its Accession to the WTO," *Regional Development Studies*, 7: 129–42.
Sit, V. and Yang, C. (1997) "Foreign-investment-induced Exo-urbanisation in the Pearl River Delta, China," *Urban Studies*, 34(4): 647–77.
Smart, J. and Smart, A. (1991) "Personal Relations and Divergent Economies: A Case Study of Hong Kong Investment in China," *International Journal of Urban and Regional Research*, 15(2): 216–33.
Solinger, D.J. (1977) *Regional Government and Political Integration in Southwest China, 1949–54*, Berkeley, CA: University of California Press.
Walder, A.G. (1995) "Local Governments as Industrial Firms: An Organizational Analysis of China's Transitional Economy," *American Journal of Sociology*, 101(2): 263–301.
Weng, J. (2006) "Study on the Structural Similarity in Industrial Sector in the Pearl River Delta," *World Regional Studies*, 15(1): 21–25.
Wu, F., Xu, J., and Yeh, A.G.O. (2007) *Urban Development in Post-Reform China: State, Market, and Space*, Oxford and New York: Routledge.
Wu, T. (2006) *Regional Planning in Modern and Contemporary China*, Beijing: Tsinghua University Press (in Chinese).
Xu, J. (2008) "Governing City Regions in China: Theoretical Discourses and Perspectives for Regional Strategic Planning," *Town Planning Review*, 79(2–3): 157–85.
Xu, J. and Yeh, A.G.O. (2003) "City Profile: Guangzhou," *Cities*, 20(5): 361–74.
—— (2005) "City Repositioning and Competitiveness Building in Regional Development: New Development Strategies of Guangzhou, China," *International Journal of Urban and Regional Research*, 29(2): 283–308.
—— (2009a) "Decoding Urban Land Governance: State Reconstruction in Contemporary Chinese Cities," *Urban Studies*, 46(3): 559–81.
—— (2009b) "Governing the Fragmented Pearl River Delta: State Reconstruction and Regional Strategic Planning," *Urban and Regional Planning Research*, 2(3): 67–83 (in Chinese).
Xu, J., Yeh, A.G.O., and Wu, F. (2009) "Land Commodification: New Land Development and Politics in China since the Late 1990s," *International Journal of Urban and Regional Research*, 33(4): 890–913.
Xu, X. (1985) "Guangzhou: China's Southern Gateway," in V. Sit (ed.) *Chinese Cities: The Growth of the Metropolis Since 1949*, Hong Kong: Oxford University Press, pp. 167–87.
Yang, B.J. (2004) "Dilemmas and Ways Out for Regional Coordinated Development in China," *City Planning Review* 81(4): 26–34 (in Chinese).
Yeh, A.G.O. (2001) "Hong Kong and the Pearl River Delta: Competition or Cooperation?" *Built Environment*, 27(2): 129–45.
Yeh, A.G.O. and Xu, J. (eds.) (forthcoming) *China's Pan-Pearl River Delta: Regional Cooperation and Development*, Hong Kong: Hong Kong University Press.

Zhang, W., Yang, S., and Gao, L. (1987) *Regional Cooperation and Economic Networks*, Beijing: Economic Science Press (in Chinese).
Zhao, Y. (2002) "From Urban Administration Towards Urban Management," *City Planning Review*, 26(11): 7–15 (in Chinese).
Zou, B. (2006) "Dilemmas and Trends of Implementing Urban Cluster Plan in China," *City Planning Review* 30(1): 1, 47–54 (in Chinese).

11 Spatial planning for urban agglomeration in the Yangtze River Delta

Chaolin Gu, Taofang Yu, Xiaoming Zhang, Chun Wang, Min Zhang, Cheng Zhang, and Lu Chen

Urban agglomeration in the Yangtze River Delta (YRD) is the engine of China's economic growth, and its solid foundation is built on economic advantages and a rich cultural background. The YRD is China's largest urban agglomeration to date, offering a vast and resourceful region for the country to carry out its globalization strategy, in order to integrate with the world economy.

Theoretical framework

The world has been transformed by globalization, regional integration, and informatization. Due to the effects of globalization, economic production has become a global, cross-boundary, and borderless activity (Mishra 1999). This facilitates and intensifies urban-networking around the world. The combined impact of globalization and integration have led to the emergence of various urban agglomerations, such as global city regions, which have become important nodes for regional development. In China, the YRD has established its role as a pioneer and a prime location for the country's integration into the global economy. Therefore spatial planning for the YRD should give sufficient consideration to globalization and its impact on urban and regional growth in the region.

The impact of globalization

Globalization, economic restructuring, and spatial reorganization have been new factors in urban development since the 1970s. Informatization, capital, technology, management, and market development have gradually spread across political, financial, and cultural boundaries, becoming freely shared across national borders. Simultaneously, new technological and economic models have created new geographical spaces. Meanwhile, the internet has gradually eliminated the traditional sense of location and distance, paving the way for the emergence of important core regions (Sassen 1991). A global, borderless economic system has begun to emerge with the advent of information technology (Yeung 2004). This implies that urban development, being affected by information flow, has become globally integrated (Castells 1996; Yeung 2004).

On a similar note, various industries have ventured into relatively low-cost areas, thereby creating a series of clusters that feature the global production chain, including small-scale enterprises in minor cities. This is possible because developed countries shift traditional industries to developing countries, while newly industrialized economies also actively relocate low-end labor-intensive industries in order to engage in capital/technology-intensive sectors (Gu 2003). From a global perspective, it has been widely noted that manufacturing industry has transferred from developed countries to East Asia and Latin America; in recent years, the manufacturing industries in East and Southeast Asia have surpassed Latin America, with East Asia particularly experiencing continued growth and development. As we look closer at China, we find that the country's eastern coastal area has become a new destination for global manufacturing industry.

Globalization and the new spatial structure

Informatization is the creation of a new "network social space," which interacts and merges with the "geographical space," thereby forming a new type of space-based interaction with the "urban space." The resulting new types of spaces comprise the global city, global city region, and the global mega-city region.

Global cities

Globalization promotes the process of internationalization in management and financing as well as in the service sectors, inevitably pushing some cities into regional or global nodes. At present, high-end service activities have become increasingly concentrated in several business cities. Sassen (1990, 1994) referred to these cities as "global cities." To date, most global cities are located in three regions—Western Europe, North America, and East Asia. In Asia, Tokyo ranks first, Hong Kong and Singapore second, and Beijing and Shanghai third. It is expected that in the near future, Beijing and Shanghai will jump into the second, or even the first tier of global cities.

The global city region

In the process of globalization, more and more people are moving from big cities to mega-cities. Some global cities are being enlarged to form global city regions. Scott (2001) believes that there are more than 300 global city regions, with the populations of at least twenty regions reaching over 10 million. In China, three major urban agglomerations in the eastern coast areas are being formed. They span different cities and provinces. According to statistical data released in 2007 on the YRD and the Pearl River Delta (PRD), the two areas combined contribute 28.09 percent to the total economic output of the country although they comprise only 1.72 percent of the territorial area and only about 10.7 percent of the total population of the country. In 2007, the total output value per capita of the two deltas was $7,000 and $10,000, respectively. These successful examples prove that China must construct other global city regions in order to reap further the benefits of globalization, and to become one of the leading global economies.

Mega-city regions

With globalization and regional integration, mega-city regions have become important economic spatial models. A global mega-city region can be defined as a city entity expanding 150 kilometers from its center. It represents a new type of urban agglomeration in which most residents live and work within a commuting distance reachable by private cars. The spatial format of a mega-city region differs because of divergence in culture and the planning system. It normally has three components.

The first component is the GBD (global business district). With the comprehensive impact of globalization and informatization, mega-cities that are of global significance accommodate financial markets, advanced service companies, banks, and headquarters of multinational corporations that, in turn, serve as drivers for urban development. Incidentally, the traditional central business districts in many cities have expanded their functional boundaries to become global business districts.

The second component is the edge city. Due to the development of ring and radiating highways in big cities, a new generation of cities has gradually emerged at the crossings. These are referred to as "edge cities," which are defined as newly extended cities that have urban functions of providing job opportunities, shopping, and entertainment to its populace (Garreau 1991; Scott 2001).

The third component is the polycentric and networked urban complex. Edge cities grow rapidly, and gradually connect with their parent cities. This results in the transformation of the traditional centripetal urban spatial structure into a polycentric pattern with low population density. These newly formed mega-city regions are generally vast, networked urban complexes that are formed by combining many cities and towns with transportation and communication systems. They have a total population of at least 20 million.

Background

In terms of the development condition, the YRD meets the requirements for building a global mega-city region. It enjoys outstanding geographical advantages in terms of proximity to the sea and the Yangtze River, and benefits from the coastal resources and inland waterways. In its inland areas, it has a developed transportation network that allows for smooth communication with the rest of the world. The YRD is also home to the biggest industries in China and is economically strong. Its automobile, steel, petrochemical, electric, information, finance, insurance, business, navigation, and telecommunications industries are considered the leading sectors. It is the largest urbanized region in China, and the most developed, culturally and technologically. To supplement its excellent industrial base, the YRD has a large network of constructed (or planned) river bridges, railways, highways, and underwater tunnels. The Shanghai and Ningbo ports rank among the top five international ports in terms of cargo throughput. Meanwhile, the Nanjing Port is the largest inland river port in China. There are airports at Pudong (International), Nanjing, Hangzhou, Ningbo, and Nantong, making the YRD an attractive destination for overseas capital. In recent years, a large amount of foreign investment has been diverted to the YRD from the PRD. In 2006, the YRD accommodated foreign investment amounting to 47.8 percent of the national total, making it the main destination for international capital transfer; this amount is greater than that invested in all of Guangdong Province in South China.

Besides the Pudong New District, the YRD contains a wide range of newly created industrial spaces, such as open coastal cities, tourist resorts, bonded zones, and science parks. This creates multiple-level and multiple-function coordination at a large geographical scale. Foreign investment has also expanded from traditional manufacturing industries into the service sectors (e.g. finance, business, real estate, and tourism).

Looking at the YRD's spatial development and its globalization process, we can identify some major problems. There is still an evident gap between the core city of Shanghai and global cities such as London, New York, and Tokyo. In 2007, the average GDP per capita in Shanghai was $8,500; for London and New York, the figure was about $30,000, and for Tokyo, over $50,000. Moreover, Shanghai still has the features of a manufacturing center. In other global cities, the proportion of tertiary industries is about 80 percent, compared to that of 51.4 percent in Shanghai. However, Shanghai and the YRD currently serve as global manufacturing centers, even if their status as financial, business, and navigation centers with great influence on the global economy have not yet been established.

Meanwhile, the construction of an international navigation center has been greatly restricted by political fragmentation in the YRD. For a long time, competition over ports among the different areas in the YRD negatively affected the realization of comprehensive port functions and the establishment of the strategic position of these ports. Such a competition likewise resulted in non-coordination between the navigation center and the other important deepwater ports, greatly restricting the economic growth and status improvement of the YRD.

There is also lack of functional complementary among different cities in the region. The common problem in the YRD is that each city aims for its own complete industrial structure, resulting in similar structures located in neighboring cities, and implicitly, a relationship dominated by competition instead of cooperation and harmony. Urban entrepreneurialism has become a priority as featured by cut-throat competition. To make things worse, the segmentation of the market and local protectionism further restrict the free flow of economic resources and dampen cooperation. This has resulted in the slow progress of regional integration and the implementation of a regional globalization strategy.

Based on the above, it could be said that there is no comprehensive regional planning and coordination in the construction of infrastructure in the region. With regard to transportation infrastructure, including ports, airports, and highways in the YRD, the duplication of construction across cities and the lack of cooperation among these have resulted in considerable deficits for some of their airports and ports. Moreover, there is no regional transportation network, further impeding regional growth. Compounding these problems is the insufficient supply of other large-scale infrastructure facilities, such as those for electricity, water supply, wastewater treatment, and communication.

The region also faces serious problems with regard to environmental pollution and an obviously deteriorating ecological environment. There are prominent issues of polluted water systems, acid rain, soil pollution, and an increasingly large amount of solid waste, and pollution issues across borders have triggered constant conflict between different cities. The excessive exploitation of underground water has resulted in the sinking of large areas of land and water pollution. Meanwhile, the deterioration of land quality has continued, and this has been made worse by a failure to protect biological diversity and ecology in the region.

Despite these problems, the YRD has a great development potential. Based on its predicted growth rate, the YRD's economy is expected to reach a total GDP of about 17,000 billion RMB in 2020, accounting for about 30 percent of the national GDP. Total population is expected to reach at least 110 million, accounting for 7.5 percent of the national total.

Currently, the total land area of the YRD is about 52,600 square kilometers, excluding mountains, forests, waterways, and parks. If calculated in line with ecological capacity, the total amount of land area available for construction in the region should not be more than half of the total land area, or about 26,300 square kilometers. The average construction land area per person for a modern city to satisfy the three basic needs of survival, development, and enjoyment should be 140 to 200 square meters. The average construction land area per person in the YRD's urban and rural communities is 140 square meters. Based on this, the capacity of land resources should not exceed that of 188 million persons. Similarly, based on the average construction area per person of 200 square meters, the carrying capacity of land resources should not exceed that of 132 million. In other words, from an ecological perspective, the maximum population in the YRD should be between 132 and 188 million.

Spatial planning

Spatial planning for urban agglomeration in the YRD mainly focuses on carrying out the globalization strategy of building the Shanghai global city, constructing the Suzhou-Wuxi-Changzhou network city, developing the Ningbo-Taizhou-Wenzhou corridor cities, and stimulating global strategic nodes. Of equal importance is improving coordination between different cities in order to make the YRD the most competitive global city region.

Shanghai as a global city

The construction of Shanghai as a global city emphasizes the readjustment of its industrial and spatial structures, realizing social transformation, and improving urban functions.

Readjusting the industrial structure

Compared to other leading cities in the Asia-Pacific Region and the rest of the world, Shanghai still lags far behind in terms of its economic size and industrial structure. In 2008, its GDP per capita was about $8,500. The tertiary sector accounted for 53.7 percent of its GDP. These figures are relatively low compared to those of global cities such as New York, London, Tokyo, and Paris. The inadequacy of Shanghai's tertiary sector will inevitably become the barrier for its future development.

As a global manufacturing and processing base, the YRD has already shown distinct industrial isomorphism, which requires Shanghai (the core city in the YRD) to develop a comprehensive producer service sector. Currently, the advanced service industry has expanded to include information technology (IT), human resource management, finance, insurance, accounting, logistics, and customer service. Shanghai, if it is to become a global city, has to develop a comprehensive service sector, especially producer services, which include insurance, business and trade, transportation and logistics, research and development (R&D), accounting, advertising, packaging, and marketing. In terms of the development of urban agglomeration in the YRD, Shanghai should also become a R&D and technological service center for the manufacturing industries.

Since 1990, the manufacturing industry in Shanghai has evolved into an advanced system. The city began to focus on high-tech industries (e.g. microelectronics, software, opto-electronics, IT, household appliances, pharmaceutical products, and new materials). These are supported by modern heavy industries (e.g. top-grade steel products, petrochemicals, fine chemicals, automobiles, and shipbuilding) and advanced equipment industries (e.g. machinery, electromechanical integrated equipment, large-set products for power stations, modern traffic equipment, refined processing facilities, specialized equipment, and energy-based equipment). The city's growth impetus will be aided by several strategic sectors, such as aviation and spacecraft, environmental protection, micro-electromechanical, new energy, and multimedia. With the development of high-tech industries, the traditional manufacturing sector has been relocated to nearby cities and distant locations in the YRD, such as Nantong and Jiaxing.

Transforming the social structure

In addition to industrial restructuring, social transformation is equally significant. There is a growing polarization of the labor force. On the one hand, rapid economic restructuring expands the wealth of a small group, such as investors and professional and technical staff. On the other hand, the unskilled labor force might lose their jobs or be forced to find other employment. Low-income families living in the city's downtown area might lose their homes due to city redevelopment or gentrification, while the demands of low-income families living in remote districts are being ignored. In a sense, developing Shanghai into a global city also requires a radical social transformation.

In general, a global city is a place where all kinds of professionals and a considerably high proportion of affluent people are concentrated. As Shanghai goes through rapid economic growth, it also faces the problem of a structural shortage of skilled professionals. For example, the proportion of skilled professionals among the total population in Shanghai is 0.51 percent, which is much lower than the 1.64 percent and 4.97 percent in the United States and Japan, respectively. Therefore it is vital for the city to attract more global talents, professionals, entrepreneurs, and investors, by improving the living, working, and cultural environments. The upcoming 2010 Expo to be held in Shanghai is a good opportunity to achieve these goals. The city should be able to realize this opportunity and produce creative, recreational, and entertaining living spaces, build a rich, fashionable, and unique cultural city image, and make the city a base for creating and trading traditional Chinese cultural products, while simultaneously developing as a consumption base for Asian cultural products.

Another dimension of social transformation is the emergence of the middle class. In a modern society, the accumulation of wealth, popularization of the concept of social justice, acceleration of social interaction, and division of power and wealth will result in a social structure that will gradually take an "olive shape." The quantity of both the wealthiest and the poorest will lessen, with the increase in the number of middle-class citizens in society. A larger number of middle-class families will be created due to diversification in ownership structure (e.g. the rise of non-state sector) and continuous upgrading of the industrial structure, because more and more people will work in new economic sectors such as finance, insurance, real estate, tourism, consultancy, and communications.

Migrant workers represent another major component in the labor force. Since 2000, the number of migrant workers seeking jobs in Shanghai has increased by 300,000 each year. According to statistics, there are currently about 4 million migrant workers in the city, of which 34 percent work in manufacturing industries, 35 percent in the service sector, and 20 percent in construction (Gu and Chen 2007). As a global city, Shanghai's prosperity will rely on this labor pool. Therefore the city must take active measures to eliminate discriminative policies, thereby creating conditions for unrestricted population mobility while accelerating the movement of the labor force. The rights of migrant workers—such as those related to social security system for household workers and a basic education system for their children—should be ensured. It is of equal importance to solve the housing shortage currently besetting these migrant workers.

The residential communities of foreigners form a new social space in Shanghai. In its fifteen-year history, the Shanghai Pudong Development Zone has attracted more than 10,000 overseas enterprises from ninety-eight countries and regions all over the world. The number of foreigners in the city as a whole has increased by about 100,000 each year. Presently, a total of 150,000 foreigners invest, work, and live on a regular basis in the Pudong District (Gu and Chen 2007). As a result, residential communities for foreigners have become an important space in the city's globalization. Some of the urgent demands in order to satisfy this requirement include those related to the living environment, culture, and lifestyle, in order to develop international communities that are deemed viable and highly attractive to foreign professionals.

As the urban social space becomes increasingly polarized, it is certain that globalization will bring about new inequalities in the urban society. Those who have capital, power, and/or professional knowledge will benefit the most from the economic transformation; those at the lower level of the society (i.e. women, new immigrants, and those without professional skills and with low educational backgrounds) may fall victim to this change. This is not only as a result of individual economic status but is also due to a divergence in social class, gender, age, race, and immigration. The construction of Shanghai as a global city must therefore take corresponding measures to deal with the drastically changing situation.

Restructuring urban space

Along with a diversified social structure, Shanghai is undergoing dramatic spatial restructuring. In recent years, informatization, highway construction, and suburbanization have contributed greatly to outward spatial expansion and inner city restructuring. Over the years, the spatial structures of many megacities in the world have changed from compact to dispersed and from monocentric to polycentric. With the goal of constructing a global city, Shanghai should have a perspective beyond the original central area of 600 square kilometers to accommodate planning for the entire municipality, which is 6,000 square kilometers in total. Furthermore, it should include the entire YRD in its spatial planning, especially focusing on the improvement of its relationship with surrounding cities (i.e. Suzhou, Jiaxing, Nantong, Ningbo, and Zhoushan). Urban structure will shift from the current model, which relies on a single center expanding outward in concentric circles, to one that features "multiple centers, networks, and openness." The new configuration will help distribute urban functions to surrounding areas, construct an ecological space around the Taihu Lake area, maintain the ecological space of the city, and create the structural framework of a giant metropolis in Shanghai.

In terms of inner city structure, planning policies will be made to encourage the development of "multiple cores, multiple centers, and multiple axes." Instead of a concentric pattern, the inner city has three functionally complementary areas: (1) downtown; (2) edge cities; and (3) new suburban regions. These three areas will form an organic inner city structure, each with divergent leading functions and development patterns. First, the central area will be developed into a global business district by integrating the Expo 2010 area into the existing CBD in Pudong and Puxi. Second, the construction of highway and railways will gradually transform the Baoshan and Minhang districts into edge cities. The latter will focus on the development of high-tech industries, with construction of the Caohejing High-Tech Park, Zizhu Science Park, and College Park. At the same time, Shanghai's image will be enhanced by the construction of a modernized Riverside New District in the Baoshan District. In this way, Shanghai will work towards becoming a global city with a top-grade steel product, and an extended industrial base, while striving towards gaining a position as a leading international shipbuilding base. It can significantly contribute to the development of the Shanghai International Navigation Center.

Third, there are two key development belts in the riverside/coastal area and in the west suburban area,[1] with economic development areas in Jiabao (Jaiding and Baoshan), Qingsong (Qingpu and Songjiang), Jinfeng (Jinshan and Fengxian), Haigang (Sea Port), Konggang (Airport), and Sandao (Chongming Island and the neighboring islands of Changxing and Hengsha). These regions are reserved areas for the future development of secondary industries and the growth of new towns. The focus is on the construction of three new cities—Lingang (near Yangshan Sea Port), Songjiang, and Jiading—where each will boast a total population of about 800,000 to 1 million. There are also the key towns of Anting, Qingpu, Jinshan, Huinan, and Chengqiao, all of which have a total population of 150,000 to 250,000. Furthermore, riverside towns are planned in the islands of Chongming, Changxing, and Hengsha Island.

To facilitate the space of flows, attention will be given to the construction of three ports (airport, sea port, and information port), two roadways (railway and highway), the Shanghai International Navigation Center, an international container hub port, an aviation hub port in the Asia-Pacific region, and a modern information port. This should include a comprehensive transportation system composed of highways, high-speed railways, and key waterways. The goal is to achieve an effective and convenient passenger and cargo transportation hub connecting domestic and overseas flows, and reaching into distant locations in the YRD.

One component of spatial planning is to form a network of open space in Shanghai. There is a need to cultivate and preserve important ecological corridors and conservation areas to achieve a multi-layer system of open space. Such a system will be composed of an ecological forest belt, open corridor areas, and wedge-shaped green areas. Moreover, there is a need to ensure harmonious development between urban and ecological spaces.

To take the above-mentioned planning policies further, it is expected that the construction of the riverside and coastal areas in Shanghai will expand from the Huangpu River to the Yangtze River, and even to the East Sea, in order to facilitate the creation of the Great Shanghai Riverside Project. Key construction projects must include provisions to develop the areas along the Huangpu River, the tourism area along the Suzhou River, the riverside space of Chongming Island, and the tourism area of the sea-spanning bridge in Dayangshan and Xiaoyangshan.

Improving city functions

The main goals of Shanghai are to develop into China's economic center, a global administration center of the manufacturing sector, a global transportation hub, and a global trading center.

To become China's economic center, Shanghai has to reinforce its status as a national economic engine and to promote its functions serving the entire country. Specifically, the aims are as follows: (1) to construct a national logistics and distribution center; (2) to reinforce the functions of Shanghai as a national financial center to serve in securities, futures, foreign exchange, and short-term private loan markets, and altogether providing various effective capital services; (3) to become a national information exchange center; and (4) to become China's international technological exchange center and platform, which will provide paid services, including two-way technology transfer and trading to customers all over China.

To become a global manufacturing administration center, Shanghai has to rely on the global manufacturing base in the YRD. Policies should be undertaken: (1) to develop a global manufacturing administration zone to provide technological services and a technological exchange platform for the manufacturing industry in the YRD and beyond; (2) to integrate the manufacturing enterprises in the YRD, and eventually the entire country, through capital operation measures, company acquisition, and reorganization—the parent companies and decision-making center should be located in Shanghai, with the branches and operations located in the YRD; (3) to make full use of its advantages as a business center, with a resourceful talent base to develop a new regional spatial structure in the manufacturing sector, using Shanghai as the "dragon head" of the region, and providing the front-end branding, marketing, and office services, while factory production is done in the YRD on the back-end; and (4) to further develop Shanghai's function as a center for market exchange, logistics, and assembly base to encourage various enterprises to establish procurement centers in the city. There is also a need to build an international manufacturing center in Shanghai. Such a process entails the transformation of the traditional manufacturing industry. By 2020, the city should become a service center, with the main focus being technological innovation and marketing in the manufacturing industry.

For Shanghai to be developed into a global transportation hub, there should be an expansion of the transshipment volume of international containers, the creation of an international container hub port by relying on the vast inlands of the YRD, and the development of an international navigation hub and global navigation center in Asia-Pacific region. Apart from hastening the construction of the Yangshan International Container Deepwater Port, the construction of the YRD International Container Hub Port (with Taicang Port in the north and the Ningbo-Beilun and Zhoushan Port in the south) should be included in the prioritized agenda. This will increase Shanghai's competitive advantage over other international ports, such as Hong Kong, Singapore, Busan, and Kobe. Moreover, there should be a comprehensive management center composed of the Shanghai Lingang New Town, the Shanghai Puxi Port Bureau, and the Shanghai Customs Clearance Center. To achieve these goals, the following must be undertaken: (1) improving the city's navigation services, creating an internationalized navigation service system, and developing the International Navigation Center in Shanghai; (2) realizing the functions of "comprehensive ports," which would be focused on Pudong International Airport and supported by the Hongqiao International Airport (making it a central airport nationally); and (3) developing an express passage railway in Pudong International Airport.

To facilitate the formation of a global trading center in Shanghai, the development of new RMB financial products should be implemented. First, the free RMB exchange service and marketization of the interest rate in Shanghai should be implemented to boost the city into becoming a global financial center. Second, Shanghai should be developed as an international business center, serving China and the Asia-Pacific region. There should be promotion of pilot projects on RMB-based products in foreign banks, as well as acceleration of the development of an offshore financial industry in Shanghai to showcase its advantages in capital, property rights, technology, and skilled professionals. Finally, Shanghai is expected to function as a sourcing center for manufacturing products, with leading ports, customs, and logistics facilities spreading to surrounding areas.

The Suzhou-Wuxi-Changzhou networked city

The strategy of city network construction for Suzhou-Wuxi-Changzhou involves integrating their urban functions, reinforcing the functional complementarities and horizontal connection among nodal cities, erecting an elastic exchange environment and a high-speed communication supporting system, and establishing guidelines to regulate market operation. These should be undertaken to maximize the interests of urban development in the Suzhou-Wuxi-Changzhou area, promote the development of Changzhou, and encourage overall economic growth in all three cities.

Specialized urban functions

Suzhou is the pioneer and leader of the YRD with regard to opening up to the world. It is the base of modern manufacturing and technological innovation. As such, it should be reconstructed to become a recipient of Shanghai's metropolitan functions and the leading nodal city in the Suzhou-Wuxi-Changzhou network. With its geographical advantage of being near Shanghai, Suzhou can serve as the nation's new high-tech industrial and technological innovation base, enhancing Shanghai's high-tech function development. This should also be the driving force behind knowledge-based activities, which include R&D, education, and technology in the networked region. Meanwhile, with a profound historical and cultural background, Suzhou's famous park resources, such as the Taihu Lake national tourist attraction, should help it transform into an international tourist city to facilitate global business and cultural exchange. The construction of the Su-Tong (Suzhou-Nantong) Bridge should be a key linkage to assist the coordinated development of both sides of the Yangtze River.

Wuxi can rely on its existing infrastructure to seize the opportunity presented by the global shift of the manufacturing industry and the rapid growth in knowledge economy. It will become an important center in the YRD by developing advanced manufacturing industries. To this end, the city should attract skilled professionals, integrate existing resources in industrial design, and film and television production, and reinforce cooperation with Shanghai, Beijing, and overseas cities in order to transform itself into a booming creative industrial center. Furthermore, Wuxi can make full use of its geographical location in the YRD to improve its transportation network and increase connectivity to surrounding areas. By constructing the Sunan International Cargo Airport in southern Jiangsu, the city is expected to become a key logistics center in the YRD.

Changzhou is another significant city. The manufacturing industry forms an essential part of its economy, which is comprised of a leading group of enterprises in machinery, metallurgy, electronics, and textiles. With globalization and changing economic conditions, the city should make use of modern technologies to reform its traditional manufacturing industries and actively develop both upstream and downstream products. It should also seek cooperation with Shanghai, Suzhou, and Wuxi in the fields of industrial development and technological innovation.

The knowledge economy

One of the leading features of a modern city is the insertion of the knowledge and culture spatial elements. The embedded knowledge economy is also manifested through urban industrial restructuring towards a knowledge-based and service-oriented economy. In the process of constructing the Suzhou-Wuxi-Changzhou city network, the focus should be on the development of the knowledge-based sector—which includes R&D, education, and innovative technologies—to achieve a Suzhou-based knowledge innovation center, improve the technology that will facilitate the development of traditional industries in Wuxi and Changzhou, and achieve the maximum growth of each city by coordinating with other cities in the network.

New and potential areas

Advancing new and potential districts as strategic spaces for the network's urban development is a necessary step for the construction of networked cities. These areas may serve as strategic nodes, focusing on key growth poles such as economic development zones, export-oriented processing areas, and reserved land areas for hub ports with global and regional strategic significance. With regard to the construction of the Suzhou-Wuxi-Changzhou city network, the leading potential areas include the Suzhou Taicang hub port area, the Wuxi Sunan International Cargo Airport area, the Changzhou Riverside Industrial Park area, and the new district of Wujin.

The Ningbo-Taizhou-Wenzhou Corridor City

The Ningbo-Taizhou-Wenzhou corridor has been one of the fastest growing and the most vibrant regions in China. The planned Hangzhou Bay Bridge and the Ningbo-Zhoushan Sea Port integration project will reinforce the relationship between this corridor and Shanghai. Its economic growth will be further closely related to Shanghai through the construction of industrial parks, a coastal passage, and a high-speed railway and port cluster. Thus the Ningbo-Taizhou-Wenzhou corridor will develop into a functional entity in southern YRD.

The development of the Ningbo-Taizhou-Wenzhou Corridor

Ningbo serves as the economic center in southern YRD and a gateway city of the Ningbo-Taizhou-Wenzhou area to connect with Pudong in Shanghai and the rest of the world. Zhoushan has an international port. This provides the potential for cooperation with Shanghai and Ningbo to develop a global hub port.

Wenzhou is an important light industry production center and export base. Its existing regional port is for domestic bulk cargo, but it has the capability to be further developed as one of the leading ports in southeast Zhejiang Province and northeast Fujian Province to support the regional economic growth of Wenzhou. Meanwhile, Taizhou will develop as an advanced manufacturing base in the YRD, and actively cooperate with Wenzhou in product research and development in order to achieve a more competitive cluster for light industry.

The corridor cities of Ningbo, Taizhou, and Wenzhou

The major problem of Ningbo-Taizhou and the Wenzhou region is the lack of a strong central city. Key functions that are best undertaken by a regional center are now scattered in different cities. This is largely due to the rapid growth of non-state economic sectors in Zhejiang Province, which has eventually led to a massive increase in specialized industries. As a consequence, different cities provide specialized functions, making it difficult to form a strong regional center. In this sense, the development of the Ningbo-Taizhou-Wenzhou corridor should follow an alternative path. Serious consideration has been given to increasing the connectivity of these cities with the core area of the YRD by promoting intercity high-speed transportation links between Shanghai, Ningbo, Taizhou, and Wenzhou, and building a coastal passage and port cluster.

Urban agglomerations in the YRD

The YRD is the most important urban agglomeration in China. In terms of regional spatial strategy, the overall objectives are: (1) to expand the territory of Shanghai as a global city to include Ningbo, Suzhou, and Nantong, modifying this cluster into a manufacturing, trading, and transportation center of the world; (2) to develop the YRD into a networked and well-connected global mega-city region, covering a total area of 30,000 square kilometers, with a total population of at least 25 million; and (3) to integrate different areas to achieve cooperative and coordinated intercity relations.

The urban system

The urban system in the YRD is relatively well developed. It is in the form of a pyramid, where cities can be divided into six tiers according to their population size. Currently, Shanghai is at the highest tier, with a population of 9.5 million in the downtown area. The second-tier cities include Nanjing, Hangzhou, and Suzhou, each with a population of about 4 million. There are ten cities at the third tier, all with a population between 1 and 3 million, including Ningbo, Wuxi, and Changzhou. Cities at the fourth tier have a population of 500,000 to 1 million, including some prefecture-level cities, such as Taizhou and Zhoushan, as well as some evolved county-level cities. Fifth-tier cities refer to medium-sized and small cities with a population of 200,000

Table 11.1 Population and hierarchy of Yangtze River Delta cities

Population (in tiers)	*Number of cities*	*City and population*	
≥5,000,000	1	Shanghai	9,500,000
3,000,000–5,000,000	3	Nanjing	4,500,000
		Hangzhou	4,550,000
		Suzhou	3,800,000
1,000,000–3,000,000	13	Ningbo	2,500,000
		Wuxi	2,000,000
		Changzhou	1,800,000
		Taizhou	1,500,000
		Wenzhou	1,500,000
		Shaoxing	1,200,000
		Nantong	1,150,000
		Zhenjiang	1,050,000
		Yangzhou	1,000,000
		Huzhou	1,000,000
		Jiading	1,000,000
		Songjiang	1,000,000
		Lingang	1,000,000
500,000–1,000,000	25	Jiaxing	800,000
		Taizhou of Jiangsu	750,000
		Zhoushan	500,000
		Taizhou of Zhejiang	650,000
		Zhangjiagang	550,000
		Jiangyin	550,000
		Yixing, Changshu, Kunshan, Wujiang, Danyang, Jingjiang, Jiashan, Pinghu, Tongxiang, Haining, Fuyang, Shangyu, Zhuji, Yuyao, Cixi, Changxing, Wenling, Linhai, Rui'an	
200,000–500,000	34	Qidong, Tongzhou, Haimen, Rugao, Rudong, Jiangyan, Hai'an, Taixing, Jiangdu, Gaoyou, Yizheng, Xinghua, Baoying, Jintan, Lishui, Jurong, Liyang, Gaochun, Taicang, Haiyan, Deqing, Lin'an, Chun'an, Jiande, Tonglu, Anji, Shengzhou, Xinchang, Ninghai, Xiangshan, Fenghua, Yuhuan, Shaoxing County, Leqing	
≤200,000	–	Yangzhong, etc.	

to 500,000 each. At the bottom tier are the medium-sized and small cities with a population of no more than 200,000 (see Table 11.1).

The adjustment of urban functions

Core areas of the YRD cover Shanghai, Nanjing, Suzhou, Hangzhou, and Ningbo.

The key urban function of these cities is focused on globalization with three policy articulations, which are as follows (see Table 11.2):

1. The urban functions of Shanghai and its surrounding areas should aim at: developing an advanced manufacturing administration and producer service center; receiving relocated world manufacturing industry while actively upgrading the industrial sector; increasing technology-intensive products and innovation capacity; exploring overseas markets; building a world manufacturing base composed of Suzhou, Wuxi, Changzhou, Nantong, Ningbo, Jiaxing, Taizhou, and Wenzhou, as well as their surrounding cities and towns; and constructing a regional port system to form a world-class navigation center through the development of the Shanghai Navigation Administration Center, the hub function of the Ningbo-Zhoushan Port, and the feeder ports of the Suzhou (Taicang), Nantong (Yangkou), Taizhou, and Wenzhou.
2. The urban function of Nanjing and its surrounding areas should aim at building a leading petrochemical industry. This can be made possible by policymaking to promote the further development of the petrochemical industry cluster in Nanjing, Zhenjiang, Yangzhou, and Taizhou, and to build the Taizhou deepwater port.
3. The urban function of Hangzhou and its surrounding areas should aim at developing a producer service sector in Hangzhou to support the growth of a world-class industrial cluster in Huzhou, Shaoxing, Jiaxing, and Haining.

Another component for improving the urban functions of the YRD cities is through regional development strategies. These include two broad policy directions. This first direction is to develop a functional hierarchy of key cities. As the YRD's core city, Shanghai serves as a national economic hub. Nanjing, the capital city of Jiangsu Province, is the central city in the north, with Hangzhou, the capital city of Zhejiang Province, being the central city in the south. Suzhou and Ningbo are two economic centers in the south and north, respectively. Wuxi, Changzhou, Zhenjiang, Yangzhou, Nantong, Taizhou, Jiaxing, Huzhou, Shaoxing, Taizhou, Wenzhou, and Zhoushan are defined as regional central cities. The second policy direction is to develop the functions of less developed regions by strengthening their linkages with key cities. Examples include Shanghai-Nantong, Suzhou-Wuxi-Changzhou, Nanjing-Zhenjiang-Yangzhou-Taizhou, Hangzhou-Jiaxing-Huzhou-Shaoxing, and

Table 11.2 Urban functions in the Yangtze River Delta

Globalization-based functional structure of YRD Cities			*Localization-based functional structure of the YRD cities*	
Subsystem	*City*	*Primary city function*	*Secondary city function*	*Functional city group*
Urban functional subsystem focusing on the global city of Shanghai	Shanghai	A global manufacturing management, traffic, and trading center; producer service center in global city region of YRD	National economic center; leading city of YRD	Shanghai, Nantong functional city group
	Nantong	International manufacturing base	Central city of Shanghai metropolitan circle's north wing; key riverside and port city; primary maritime industrial base in YRD's north wing	
	Suzhou	Advanced global manufacturing base, technological innovation center in YRD's global city region	Sub-central city in Shanghai metropolitan circle; economic center of YRD's north wing	Suzhou, Wuxi, Changzhou, Taixing functional city group
	Wuxi	Advanced international manufacturing base	Regional central city; creative industry base; regional traffic hub; tourist destination	
	Changzhou	International manufacturing base	Regional central city; special industrial base, tourist destination	
	Taizhou	Extended area of international manufacturing base	Regional central city; key riverside port city	

Table 11.2 Continued

Globalization-based functional structure of YRD Cities			*Localization-based functional structure of the YRD cities*	
Subsystem	*City*	*Primary city function*	*Secondary city function*	*Functional city group*
	Ningbo	Important part of Shanghai International Navigation Center; service base focusing on international business; advanced manufacturing base focusing on petrol-chemical and auto industries	Economic center in YRD's south wing; key energy and raw material bases; important foreign trade port in Eastern China	Ningbo (Zhoushan), Taizhou, Wenzhou functional city group
	Zhoushan	Important part of Shanghai International Navigation Center	Garden city in YRD featuring maritime industry	
	Taizhou	International manufacturing base	Modern port city in coastal area of Southeast China	
	Wenzhou	International light industry production and export base	Modern port city in coastal area of Southeast China	
	Jiaxing	International manufacturing base	Central city in Shanghai metropolitan circle's south wing; central city in the areas around Hangzhou Bay; cultural and tourist city	Hangzhou, Jiaxing, Huzhou, Shaoxing functional city group
Urban functional subsystem focusing on Hangzhou	Hangzhou	Sub-center of producer service sector in YRD's global city region; international tourist destination	National historic and cultural city; central city in YRD's south wing; information innovation center and information hub	

Table 11.2 Continued

Globalization-based functional structure of YRD Cities			*Localization-based functional structure of the YRD cities*	
Subsystem	*City*	*Primary city function*	*Secondary city function*	*Functional city group*
	Huzhou	International manufacturing base	Important nodal city in YRD Area; central city in Hangzhou-Nanjing Development Belt Zone; connecting area among Zhejiang, Jiangsu, and Anhui Provinces	
	Shaoxing	International manufacturing base	Advanced manufacturing base focusing on textiles; recreational and tourist city in YRD's south wing	
Urban functional subsystem focusing on Nanjing	Nanjing	Global petrochemical industry base; sub-center of producer service sector in global city region in YRD	Famous national historic and cultural city; central city in YRD's north wing; regional traffic center; regional education and research base	Ningbo, Zhenjiang, Yangzhou, Taixing functional city group
	Zhenjiang	International petrochemical industry base	Important energy base in YRD; regional logistical activity center; ecological city for human habitats and business	
	Yangzhou	International petrochemical industry base; international tourist destination	Famous national historic and cultural city; regional central city	

Ningbo-Taizhou-Wenzhou. Nantong has recently been partnered with Shanghai to receive relocated manufacturing industry, which facilitates the functional incorporation of Nantong in the Shanghai megalopolis. With the construction of the Ningbo-Taizhou-Wenzhou express highway, the less developed regions will soon become an important part of the YRD urban system.

Regional coordination and division of labor

In order to enhance the competitiveness of the YRD, coordinated regional development is essential. To this end, it is planned that the YRD cities should develop a division of labor, to avoid excessive competition. Shanghai will focus on advanced service sectors, such as finance, trade, shipping, and information service. It will also stress the development of durable commodity industries (e.g. automobiles and personal computers) and high-tech industries. The heavy and chemical industries will maintain or slightly increase their current capacities. North and South YRD will focus on developing light industries (e.g. textile, durable commodities, machinery, and petrochemical industries) and industrial specialization, while encouraging the growth of the tertiary sector, such as tourism.

Strategic areas under globalization

Some global strategic areas are planned to facilitate China's engagement with the global economy. These areas are new urban growth nodes, which are functionally similar to the economic and technological development zones launched in the 1980s, and the high-tech development zones introduced in the 1990s. However, they have a higher degree of development intensity and openness compared to the preceding decades. In order to construct global strategic areas, the state should establish corresponding policies, laws, and regulations to promote the development of the "world factory" in the southeastern coastal region and the growth of a global shipping and logistics hub.

In the YRD, the aims should be to develop global strategic areas focusing on port-related industries, such as trade and logistical service, and on the advanced manufacturing sector. Those port-related strategic spaces will become gateways to the international market. They should be allowed to adopt policies and operation mechanisms similar to those of free trade areas to provide offshore financial business integration with their port functions and the high-end logistics sector. Those strategic areas of advanced manufacturing industries should be encouraged to closely engage with the global economy and become pilot fields, using high technology to upgrade the traditional manufacturing sector. Through the construction of global strategic areas, globalization in the YRD will be facilitated, promoting and progressing the national globalization strategy. Spatially, global strategic areas will be located at places with favorable regional conditions, good development potential, and strategic significance. These areas include the following:

1. *Shanghai New Port City.* Located within the Shanghai Nanhui District and on the west end of the East Sea Bridge, the new port city has a total planned area of 50 square kilometers. The aims are to rely on the Yangshan Deepwater Port, the Ningbo-Zhoushan International Hub Port, and the Pudong International Aviation Hub to develop a modern navigation service, international trade, and advanced logistics sectors. This area will be transformed into an integral part of the international navigation center in the YRD.
2. *Hangzhou-Jiangdong New Town.* This zone will make use of the reclaimed land along the river in the Hangzhou-Xiaoshan District, with a total planned area of 30 square kilometers. The main projects will include an Asian packaging center, the renovation of the Xiaoshan International Cargo Airport, and the development of a logistical center between the airport and Jiangdong Industrial Park. It will be formulated into a service base to help small enterprises in Wenzhou and Taizhou to establish international markets and to improve the added-value of their products. It will become a global strategic area to highlight the functions of the airport and seaport.
3. *The Suzhou-Taicang Port Zone.* This site is located at the mouth of the Taicang Liuhe River, with a total planned area of 30 square kilometers. It is an important deepwater container port in the inland area along the Yangtze River. The focus will be on the construction of fourth- and fifth-generation container terminals, and the development of the port logistics sector. The aim is that it will become one of the most important segments of the International Navigation Center hub ports in the YRD.
4. *The Nantong-Yangkou Port Area.* This is a strategic site located within the Yangkou Port Economic Development Zone, with a total planned area of 20 square kilometers. It has a favorable deepwater condition due to its proximity to the Yangtze River, allowing for the development of a robust logistics industry. Berths for 100,000-ton and 200,000-ton vessels will be constructed, and an intermodal transportation system of railways, highways, inland waterways, and canals will be built to make it an important specialized port and port logistics base in the north wing of the International Navigation Center in the YRD.
5. *The Taizhou-Jingjiang Port Area.* This is a site located on the north end of the Jiangyin Bridge and Bawei Port Area, with a total planned area of 20 square kilometers. The focus will be on creating a regional logistics, energy, and metallurgy industrial base in the YRD. It is a rising strategic area that can take on the relocation of the global manufacturing industry. The emphasis must be on realizing the cross-river development in North YRD, and serving as the pioneer in the joint progress of north and south Jiangsu Province.
6. *Ningbo-Hangzhou Bay Area.* This strategic site is located on the south end of the Hangzhou Bay Bridge in the city of Cixi, with a planned area of 30 square kilometers. The initial development has already been realized;

the area currently has a relatively satisfactory infrastructure and an adequate land reserve. The construction of the sea-crossing bridge along the Hangzhou Bay facilitates communication with the global city of Shanghai. The area should maximize its advantages, including the strong automobile and industrial cluster, to focus on the development of an independent brand of auto manufacturing. This will encourage the growth of local medium-sized and small enterprises, allowing them to participate in global competition, and receive relocated advanced manufacturing industries from overseas countries.

7. *The Nanjing Chemical Industry Park.* This site is located in the Nanjing Luhe District, north of the Second Yangtze Bridge, with a total planned area of 50 square kilometers. The aims are: to construct deep-processing products with high added-value, to make it one of the leading chemical industry parks in China, with global influence; to emphasize the development of the ethene, acetic acid, and chlor-alkali industrial chains; and to promote the development of fine chemistry, pharmaceutical chemistry, high molecular material, and other downstream-supporting and deep-processing projects. The completed Yangtze Petrochemical, the integrated Yangba project, and the adjacent Jinling Petrochemical, Nanjing Chemical Fiber, and Yizheng Chemical Fiber, will form a globalized riverside petrochemical industry belt, with the aid of a modern road and pipe network.
8. *The Changzhou Riverside Industrial Park.* This is a site located in the Changzhou Xinbei District close to the Yangtze River, with a total planned area of 30 square kilometers. The focus is on developing equipment manufacturing, pharmaceutical chemistry, new materials, and other advanced manufacturing industries. The main objective should be to revive the economy of Changzhou and to make it an advanced manufacturing production center in the YRD. This will also contribute to the important functions of a globally advanced manufacturing center in the YRD, along with the industrial innovation center in Suzhou, and the international cargo airport and logistics center in Wuxi.

The global mega-city region

The mega-city region in the YRD is composed of inner, middle, and outer layers. The inner layer encompasses the Shanghai-Nanjing-Hangzhou Urban Cluster, the middle layers cover the hinterland of Nanjing and Ningbo, and the outer layer consists of a series of peripheral cities surrounding the Shanghai-Nanjing-Hangzhou Urban Cluster (see Figures 11.1 and 11.2).

Spatial structure

The global mega-city region in the YRD comprises Shanghai and other secondary cities, including Nanjing, Suzhou, Hangzhou, and Ningbo. Together,

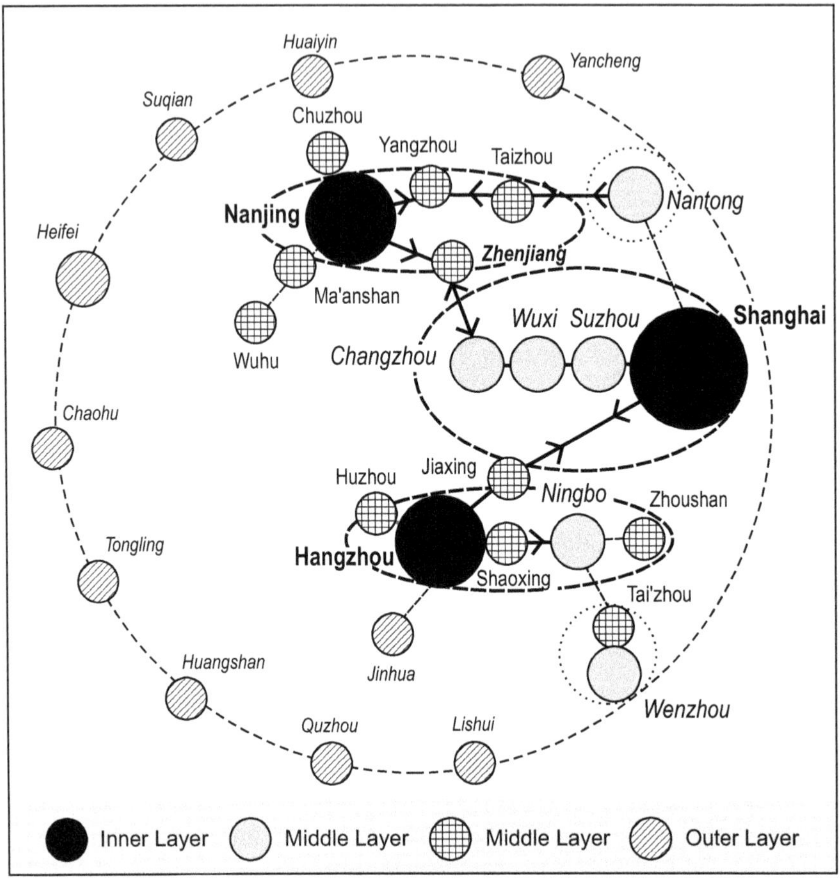

Figure 11.1 Spatial levels of the mega-city region in the Yangtze River Delta

Source: Chapter authors.

they form the first-level urban system. Wuxi, Changzhou, Nantong, Taizhou, Yangzhou, Zhenjiang, Jiaxing, Huzhou, Shaoxing, Zhoushan, Taizhou, and Wenzhou comprise the second-level system; while the county level cities, such as Jiangyin, Cixi, and Danyang, are the third-level system (see Figure 11.3).

The Shanghai global city megalopolis covers part of the central areas of Nantong, Suzhou, Jiaxing, and Ningbo. The sub-region, centered on Nanjing, covers two central cities at the prefecture level, (i.e. Yangzhou and Zhenjiang), as well as the county cities of Jiangdu and Yangzhong. The sub-region around Hangzhou adjoins three prefecture centers, namely Shaoxing, Huzhou, and Jiaxing, as well as the county cities of Tongxiang and Haining. The sub-region around Suzhou reaches the cities of Wuxi, Changzhou, Taicang, and Kunshan. The sub-region centered on Ningbo overlaps Taizhou of Zhejiang, Wenzhou, and Cixi.

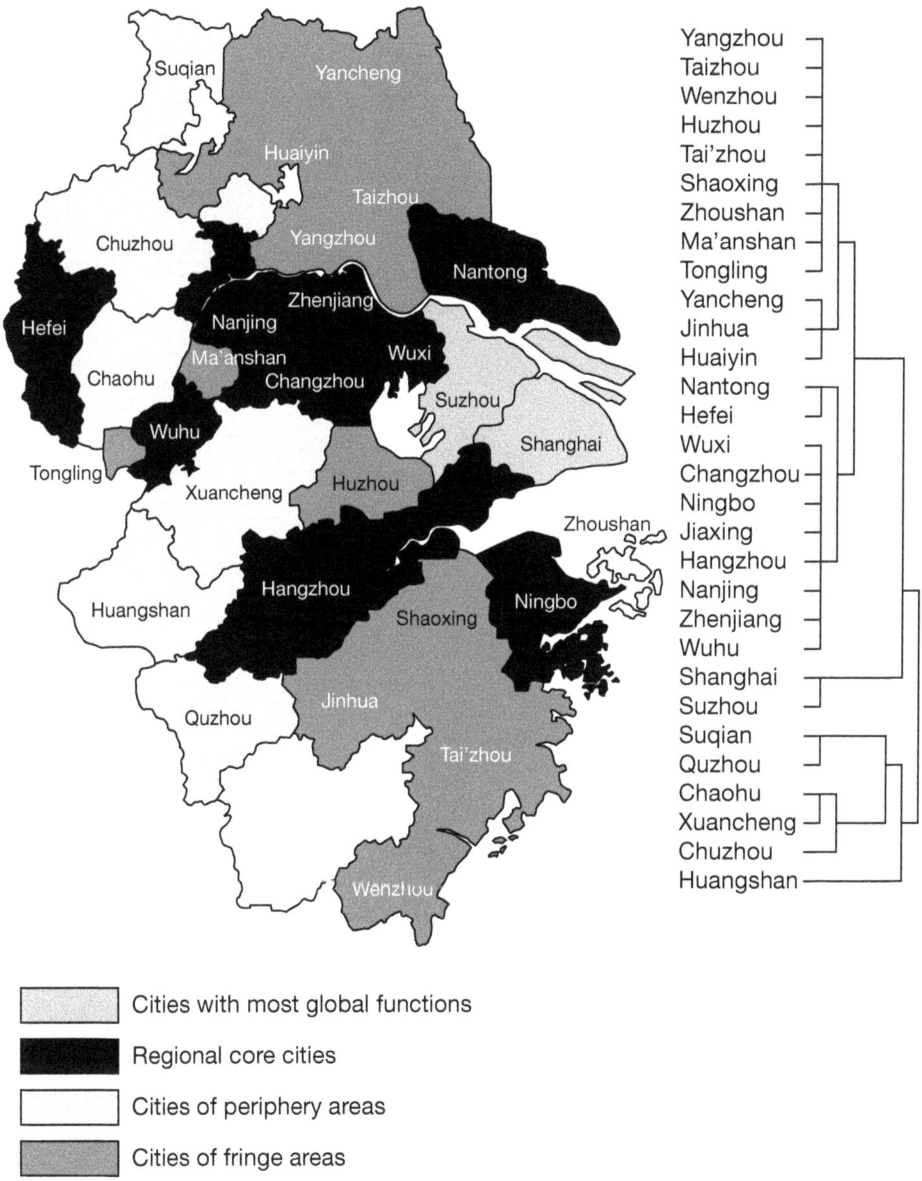

Figure 11.2 Spatial structure of the mega-city region in the Yangtze River Delta

Source: Chapter authors.

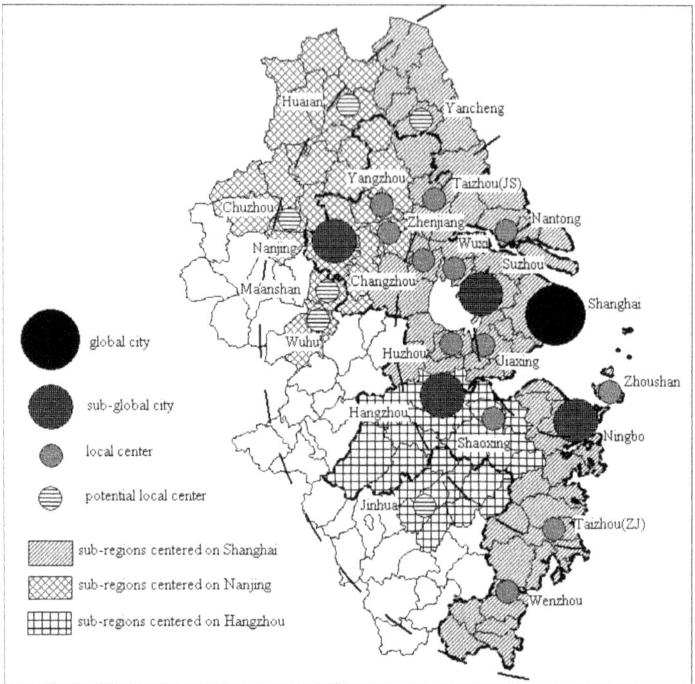

Figure 11.3 Spatial structure planning of the global city region in the Yangtze River Delta

Source: Chapter authors.

Organization of urban function

The YRD's global mega-city region can be divided into three functional regions (Shanghai, Nanjing, and Hangzhou) and two sub-regions around Suzhou and Ningbo (see Figure 11.4).

The Shanghai region consists of four hierarchical areas. The first is the core area, commonly referred to as the downtown of Shanghai. The second is the Shanghai metropolis, sweeping the new towns of Chongming, Jiading, Songjiang, and Lingang (near the Yangshan Seaport). The third, which is referred to as Shanghai megalopolis, includes the central cities of Nantong, Suzhou, Jiaxing, and Ningbo. Finally, the global city region of Shanghai incorporates some front-edge zones bordering Taizhou, Changzhou, Huzhou, and Wenzhou.

Supporting systems

The leading supporting systems of the YRD's mega-city region involve an express communication system, a regional transportation-logistics system, information infrastructure, and ecological, environmental, and energy protection systems.

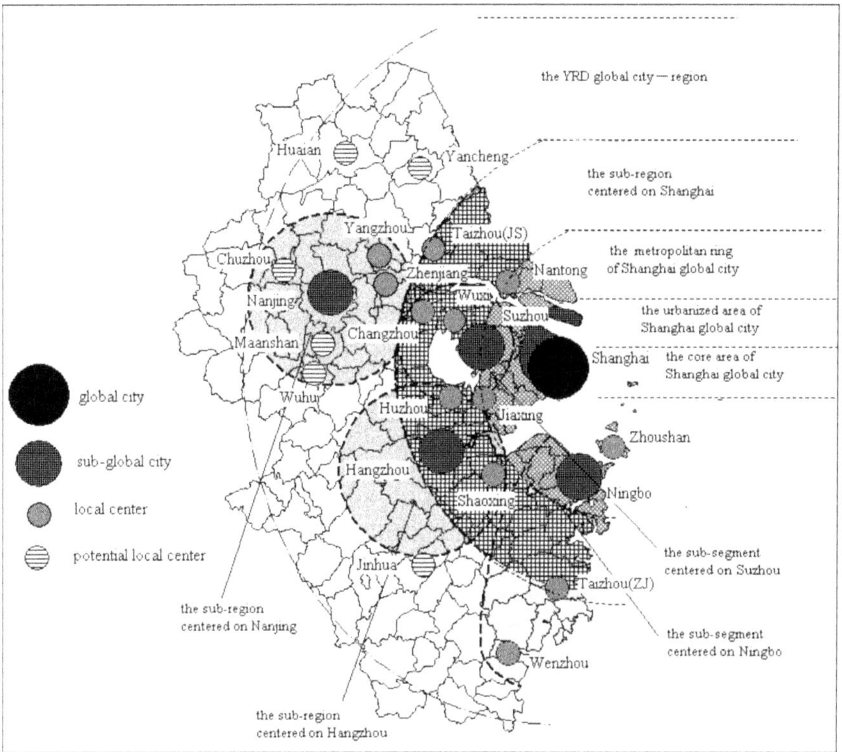

Figure 11.4 Structure of the multi-centered global city region in the Yangtze River Delta

Source: Chapter authors.

Express communication system

To develop the YRD into a global city region in conjunction with Shanghai as a global city, the first mission is to construct an international aviation hub system and advance an international aviation network linking global cities around the world in order to reinforce ties with the global urban system. The aviation system should be hierarchical, including an international passenger aviation hub, a cargo hub, and feeder airports. Major aviation hubs include a passenger aviation hub in Pudong and two cargo hubs in southern Jiangsu and southern Zhejiang. Moreover, Nanjing-Lukou International Airport should have more international passenger and cargo routes. The construction of a cargo base should be emphasized, in order to improve cargo handling and distributing capacity. Furthermore, the feeder airports in Hangzhou, Xiaoshan, Ningbo, Zhoushan, Taizhou, Wenzhou, Nantong, Changzhou, and Suzhong (under construction), should also be further promoted.

From the perspective of regional sustainable development, the construction of new large-scale highways is not recommended; rather, the focus should be on the formulation of a highway commuting system between major metropolitan areas, especially a "one-hour" highway commuting circle or something similar. In the future, in order to accommodate rapid economic growth and the integration of the region, emphasis should be placed on the erection of a high-speed railway and intercity rail transit systems.

Transportation-logistics system

There are four transportation-logistics sub-systems, which are summarized as follows:

1. *International shipping center.* The mega-container hub port should mainly involve the port areas of Waigaoqiao, Dayangshan, Xiaoyangshan, Ningbo, Zhoushan, and Taicang. Through the reorganization of their functions and coordinated development, they will jointly form the international container hub port in the YRD. An international navigation management center should be established to integrate and coordinate the main functions of the Shanghai Lingang New City, the Shanghai Puxi Port Administration Bureau, and the Shanghai Customs Clearance Center.
2. *Free trade zone.* The development of the YRD must focus on the following: the construction of the Pudong, Zhoushan, and Rudong free trade zones; the promotion of the transshipment volume in these zones; the construction of information systems for ports and logistics; and the improvement of port-related industry and the service sector. The Pudong free trade zone should promote the interaction between the Waigaoqiao and Yangshan ports. The Zhoushan free trade zone has two areas supporting the Shanghai-Zhoushan and Ningbo-Zhoushan port clusters. To promote the growth of the Rudong free trade zone in Nantong, emphasis should also be accorded to the use of the radiating effect of the Shanghai navigation center on Yangkou port and the construction of the Shanghai-Chongming-Suzhou river-crossing passage. Moreover, emphasis should also be placed on the construction of the ports located in Ningbo-Zhoushan and Shanghai-Zhoushan, as well as the deepwater port in Jiangsu Rudong.
3. *Container feeder ports.* The feeder port cluster in the YRD revolves around the Shanghai port, and is divided into the south and the north wings in Zhejiang and Jiangsu provinces, respectively. The focus in the north is the construction of feeder ports in Nanjing, Nantong, Zhenjiang, Taicang, and Jiangyin, while the focus in the south is the construction of supporting ports for the navigation center, such as the Jiaxing, Taizhou, and Wenzhou ports. The latter two ports should enjoy additional advantages from this, and should be integrated and developed jointly with the Damaiyu deepwater port area.
4. *Express port transportation system.* An effective network of express port transportation systems will be of great significance to the development of a hub port. Such a network should incorporate Dayangshan-Xiaoyangshan,

Ningbo-Zhoushan, and Shanghai-Zhoushan cross-sea bridges, the coastal railway, and highways and express highways linking various ports.

Regional transportation network

For a long time, cities in the south wing of the Yangtze River, such as Ningbo and Taizhou, have had a weak link with Shanghai due to the presence of Hangzhou Bay. The construction of a cross-sea bridge will enhance the connection and facilitate the formation of urban agglomeration in the YRD. Moreover, Shaoxing and Xiaoshan have also begun planning the construction of the Hangzhou Bay Bridge, while Zhoushan is currently planning the construction of a cross-sea passage to connect with the East Sea Bridge. This will greatly reduce the spatial distance between Shanghai and the coastal areas in southeast Zhejiang Province, merging the south wing of the YRD with the Shanghai economic circle.

To take this further, the construction of river-crossing passages will benefit the spatial reorganization of the two sides of Yangtze River. Currently, there are three river-crossing passages, namely, the Nanjing Yangtze River Bridge, the Jiangyin Bridge, and the second Nanjing Yangtze River Bridge. However, an increase in the number of river-crossing passages remains a key and basic component and guarantees the implementation of a development strategy along the river in Jiangsu Province. These projects will also reinforce the linkages between Shanghai and the southern and middle areas of Jiangsu Province. Together with the Su-Jia-Hang (Suzhou-Jiaxing-Hangzhou) Highway, the Lian-Yan (Lianyungang-Yancheng) Highway and other truck routes, these river-crossing passages will help form a network of transportation links radiating from Shanghai.

Information infrastructure

Information infrastructure, such as the information portal for financial services and its technical supporting system, should be improved. This will benefit the R&D sector and the intensive application of advanced technologies, including the new generation of mobile communication, soft switch, and intelligent optical networks. The service capacity of such infrastructure should be greatly enhanced to provide a high-performance information platform. In this way, the YRD will develop into one of the major information hubs in the Asia-Pacific region.

Ecological and environmental protection

Previously, the YRD's development has mainly focused on economic growth. This growth-driven pattern has led to disastrous environmental consequences. In order to achieve regional sustainable development, all cities have planned and developed a number of megaprojects to protect the environment. Moreover, the socio-economic integration in the YRD has facilitated regional cooperation in environmental protection. Many megaprojects for ecological conservation have emerged, such as the Taihu Lake Treatment and Conservation, the Inland Water System Treatment and Conservation, and the River Ecological Projects.

Energy supply

Rapid socio-economic growth in the YRD has generated a huge demand for energy. Policies should be made to explore a variety of energy sources from domestic and international markets. Renewable energy and resources should also be explored and developed to replace traditional energy sources, such as coal and oil, and a group of megaprojects are planned in the YRD to secure more energy sources for the region. It was argued in previous research that there is an urgent need to explore new power sources, such as nuclear power and natural gas. For instance, policies will be formulated to make full use of the West–East natural gas transmission project to encourage large-scale application of natural gas. This effort can be taken further by actively exploiting the natural gas reserve in the sea. A large-scale project by Ningbo is the development of the Chunxiao Natural Gas Field, which lies in the East China Sea.

Acknowledgments

The chapter was based on the research report "Planning for the Development of Urban Agglomeration in the Yangtze River Delta." The project was sponsored by the Department of Development Planning of China's National Development and Reform Commission (NDRC). The researchers obtained help from the following organizations: the Department of Development Planning of the NDRC; the Urban Development and Reform Commission of Shanghai, Jiangsu Province, Zhejiang Province, Nanjing, Changzhou, Hangzhou, and Ningbo; the Zhejiang Province Construction Bureau; the Urban Planning Bureau in Shanghai, Changzhou, Hangzhou, and Ningbo; the Urban Planning and Design Institute in Zhejiang Province, Changzhou, Hangzhou, and Ningbo; the Shanghai Academy of Social Sciences; the City and Regional Research Center of East China Normal University; and the Department of Regional and Urban Planning in Zhejiang University. We hereby express our sincerest gratitude to them for the help. This chapter is a part of the Key Project of the Natural Science Foundation (40435013) and of the Key Subjects of PhD Station Foundation, launched by the State Ministry of Education.

Note

1 The coastal development zone involves Pudong, Nanhui, Fengxian, and Jinshan, while the western suburban area includes Baoshan, Jiading, Qingpu, and Songjiang.

References

Castells, M. (1996) *The Rise of the Network Society*, Oxford: Blackwell.

Garreau, J. (1991) *Edge City: Life on the New Frontier*, New York: Doubleday.

Gu, C. (2003) "Urbanization Studies: An International Approach," *City Planning Review*, 27(6): 19–24 (in Chinese).

Gu, C. and Chen, L. (2007) "Construction of Shanghai as a Global City from the Perspective of City Cluster in the Yangtze River Delta," *Areal Research and Development*, 26(1): 1–5 (in Chinese).

Mishra, R. (1999) *Globalization and the Welfare State*, Cheltenham: Edward Elgar Publishing.

Sassen, S. (1990) "US Immigration Policy toward Mexico in a Global Economy," *Journal of International Affairs*, 43(2): 369–83.

—— (1991) *The Global City: New York, London, Tokyo*, Princeton, NJ: Princeton University Press.

—— (1994) *Cities in a World Economy*, Thousand Oaks, CA: Pine Forge/Sage Press.

Scott, A.J. (ed.) (2001) *Global City-Regions: Trends, Theory, Policy*, Oxford: Oxford University Press.

Yeung, Y.M. (2004) *Cities in Asia and Pacific Region Under Globalization*, Beijing: Science Press (in Chinese).

Index

For Product Safety Concerns and Information please contact our EU
representative GPSR@taylorandfrancis.com
Taylor & Francis Verlag GmbH, Kaufingerstraße 24, 80331 München, Germany

www.ingramcontent.com/pod-product-compliance
Ingram Content Group UK Ltd.
Pitfield, Milton Keynes, MK11 3LW, UK
UKHW022220190926
475702UK00003B/461
9780415560894